BLUEPRINT FOR GOING GREEN

# BLUEPRINT FOR GOING GREEN

How a Small Foundation Changed the
Model for Environmental Conservation

Gerald P. McCarthy

University of Virginia Press • *Charlottesville and London*

University of Virginia Press
© 2024 by Gerald P. McCarthy
All rights reserved
Printed in the United States of America on acid-free paper

*First published 2024*

ISBN 978-0-8139-5072-3 (hardcover)

ISBN 978-0-8139-5073-0 (ebook)

9 8 7 6 5 4 3 2 1

Library of Congress Cataloging-in-Publication Data is available for this title.

*Cover design:* Kelley Galbreath

*For Judge Robert R. Merhige Jr. (1919–2005),*
*Federal District Court, Eastern District of Virginia,*
*who set this story in motion*

# CONTENTS

# INTRODUCTION

"ON FEBRUARY 1, 1977, a federal judge did something that no court had ever done before. He turned a fine for pollution into a creative way to benefit the people and the environment of Virginia."[1] Instead of following routine and sending the fine, which was the largest federal fine ever imposed for a water pollution violation, to the United States Treasury, Judge Robert R. Merhige Jr. caused the creation of the Virginia Environmental Endowment, a private nonprofit grant-making organization, to improve the quality of Virginia's environment. The Endowment's work since 1977 has resulted in land conservation, tangible improvements in water quality, and a network of environmental organizations to advocate for improving and enforcing environmental protection.

In that ruling, which was about polluting the James River with the insecticide Kepone, the worst environmental disaster in Virginia's history, Judge Merhige held Allied Chemical Corporation accountable with a $13.24 million fine ($62 million today) "because that's the maximum the law allows me." Judge Merhige's decision struck like a bolt of lightning, because up until that case, polluters had been lightly regulated, rarely called to account, and free to discharge all manner of waste without disclosing its contents. It was front-page news from Washington, DC, to Richmond to Norfolk.

What is more, instead of requiring the fine to be paid in the usual manner, to the federal government, he encouraged Allied to develop a way for it to be used to benefit Virginians. Allied agreed to make a voluntary payment of $8 million to start an environmental fund for Virginia. Judge Merhige then reduced Allied's federal fine by the same amount. Allied still paid out a total of $13.24 million, but $8 million ($38.95 million in today's currency) was used to establish the Virginia Environmental Endowment. What happened, how we brought this idea to life, is the story that unfolds in this book. It is a story filled with people whose long-term dedication in the face of obstacles large and small has made a significant difference in the quality of Virginia's environment and in the lives of Virginians. This story is about the development of the modern environmental movement

1

in Virginia and how the Virginia Environmental Endowment helped to accelerate that development.

The idea of turning a fine for pollution into a creative way of helping to address environmental problems was unprecedented. To appreciate how significant this decision was, it is necessary to recall that federal laws regulating air and water pollution were still new in 1977. There were no "friends of the river" groups. There was no Chesapeake Bay Program, nor citizen participation in air and water regulatory decisions. There was almost no private land conservation program. Many of these developments came into being with financial help from the Virginia Environmental Endowment. The decision set the stage for an organization that would become a major force for improving the quality of Virginia's environment, though few of us at the time imagined or foresaw the possibilities.

It was Rachel Carson's 1962 book, *Silent Spring,* that first sounded the alarm about toxic substances hurting the environment. Carson's book argued that pesticides were harmful to the environment. The book raised the visibility of environmental concerns, particularly the idea that these chemicals might have negative side effects on the environment. It would take years before this general idea of chemical harm to health and the environment became clear. For the purposes of this book, the significance of the Kepone story is that it brought further focus to the problem of unnoticed toxic wastes being discharged into the air and water.

Beginning not with the disaster itself but rather with what happened next, this book shows positive examples of how to improve the environment. It is not a woe-is-me, "the world is going to hell," negative rant. It is about the development of the modern environmental movement in Virginia and how the Virginia Environmental Endowment helped to accelerate that development. People want hope and encouragement, because people are getting together and telling their elected representatives that they want a clean environment and enough taxpayer money dedicated to make sure it stays that way.

The Endowment's grant-making over the next few decades turned the judge's idea into a world of good for the environment. A key factor in the Endowment's success has been the requirement of matching funds for each grant, a practice that has gotten more organizations interested in its work and, in turn, gotten more people and more money involved in improving the environment in all corners of Virginia. During the thirty-six years of my tenure as director, 1977 through 2013, VEE made grants totaling about $28 million, which when combined with matching funds represented an investment of almost $70 million in environmental improvement.

In the decades since 1977, tremendous progress has been made one step at a time—sometimes two steps forward and one step back, depending on how the political wind socks were blowing. Overall, though, the environment is in much better condition, there are stronger laws in place, and perhaps most importantly, there are many nonprofit public-advocacy groups today to hold polluters and regulators accountable. Citizens, aided by the law, have defeated attempts to weaken environmental regulations and will continue to do so. A new generation of environmental leaders is emerging.

The Endowment helped thousands of people protect clean water, conserve landscapes, advance the science of fisheries management, and influence public policy and state funding for natural resources. VEE funded paradigm-shifting research in Chesapeake Bay fisheries management and, through a new partnership between business and conservation, encouraged the state to invest hundreds of millions of dollars to protect water quality and conserve land throughout the Commonwealth—in accordance with its laws. In essence, VEE acted as venture capital for environmental improvement in Virginia. We eventually leveraged those funds into hundreds of millions of new dollars for water quality improvements and land conservation in Virginia's budget.

Also crucial has been VEE's emphasis on research. VEE made grants for scientific research in support of public policies, legal research, and analysis to improve laws and even surveyed public opinion to gauge the public's attitude about environmental problems. For example, at a critical moment when the national government was pursuing agendas that would weaken established environmental laws—asserting that environmental regulations are "costly, burdensome, and job-killing," casting polluters as the victims of laws and regulations designed to improve the environment—VEE commissioned the first-ever public poll documenting the Virginia public's attitudes about the environment. The results showed clearly that most voters wanted a clean environment as well as laws and regulations to enforce that goal. It also gave new confidence to the conservation groups that their work was needed, valued, and appreciated.

VEE's autonomy has proved critical as well. When the judge created the Endowment, he gave the board complete independence to do what it thought best to conduct its mission. They understood and appreciated this freedom to act independently and were entrepreneurial, strategic, and focused. Given the origin of the Endowment in the Kepone disaster, the board took direct aim at identifying opportunities to prevent future "Kepones" from happening. When we looked at the subject of "toxics and water quality," for example, we made grants to different organizations with complementary expertise in science, public policy, and law. However, the more we learned, the more we realized that public participation and advocacy were also a necessary—and largely missing—component of efforts to improve the environment in 1977. Soon after, therefore, providing initial funding for environmental nonprofits became a key part of our grant-making.

When the Endowment began operating in 1977, it was the only grant-making organization in the country that focused exclusively on environmental quality as its mission. Indeed, in Virginia the Endowment was the only conservation-focused funder for quite a while. In addition, there was, for example, no public participation in the development and implementation of laws and policies, participation that, thanks in part to VEE's work over the years, we now take for granted. It is hard to imagine now, but in 1977 there were hardly any advocacy groups in the state insisting that Virginia care for its environment: no Southern Environmental Law Center; no Elizabeth River Project; no Chesapeake Bay Foundation office in Virginia, nor a Chesapeake Bay Program; no Virginia Conservation Network; no Natural Heritage Program; no environmental mediation center. The Valley Conservation Council hadn't been formed, and the local land trust movement had not yet taken hold in Virginia. Nor had the Environment Virginia Symposium been established, an annual event where business, government, and conservation organizations could meet and exchange ideas each year. The Endowment helped to start all these programs in Virginia and many more. Their stories show who was really responsible for the progress made over a long period of time: Virginians.

THIS IS a book filled with positive examples of how people can improve their environment, with ample reasons for hope that a cleaner environment is not only desirable but possible. It also provides a template for how even the smallest foundations can leverage their limited funds to catalyze social innovation and community improvement by focusing their efforts, targeting their grants, and persisting over time. It is a road map of how people can replicate Virginia's best efforts by focusing on results.

Most of the time, change happens slowly and incrementally. If you were to revisit this country during the late 1960s, you could hardly fail to notice—for example—that the Cuyahoga River had caught fire again, that a major oil spill had fouled the coast of California, and that the air in Los Angeles and New York City was so dirty you could see it. Many cities and towns dumped raw, untreated sewage right into their adjacent rivers instead of valuing and embracing the economic possibilities they represented.

When I came to Virginia in 1970, environmental quality was a popular, bipartisan cause. Only in the political sphere has that common ground broken down. VEE's first poll in 1995 clearly showed the divide between the people and the politicians over the environment, which became an inflection point in the fight against pollution both in Virginia and nationally. Virginians want a clean environment, and they know that there is no such thing as a right to pollute. Indeed, the constitution of Virginia asserts the opposite: a mandate for the state to protect and conserve the natural environment.

The new Virginia constitution adopted in 1971 stated in Article XI that it is Virginia's policy to "protect its atmosphere, lands, and waters from pollution, impairment, or destruction." In this spirit, armed with conservation as a fundamental public policy and enabled by financing from the Virginia Environmental Endowment, the environmental movement in Virginia has overcome the skepticism of politicians, the inertia of environmental bureaucracies, and the constant challenges of raising money, becoming an effective advocate for better laws, and their enforcement, to protect and conserve the environment.

As we will show, leadership emerges out of a sense of place and a mission to right a wrong, and it has taken many forms. One governor turned a fledgling Chesapeake Bay Program into a powerhouse for preserving the Bay. One environmental lawyer—a new specialty developed over the last fifty years—turned his dream of a clean environment into a multistate public interest law firm headquartered in Virginia and operating throughout the South.

Perhaps the Endowment's greatest impact is the financial help VEE provided to launch and sustain so many citizens' nonprofit organizations. These groups have led the way to environmental improvement by demanding that Virginia's elected officials carry out their constitutional duty, as stated in Article XI, to "protect its atmosphere, lands and waters from pollution, impairment or destruction for the benefit, enjoyment and general welfare of the people of the Commonwealth."

Over the past few decades, citizens have organized local environmental groups to protect their special places. The Valley Conservation Council in the beautiful Shenandoah Valley of Virginia has organized the voluntary and permanent protection of thousands of acres, using Virginia's generous tax credit for permanent easement protection of such property. Citizens upset with the water pollution they found in their neighborhoods developed a network of "friends of the river" groups, from the Shenandoah to the Rappahannock, the James, the Elizabeth, the Lynnhaven, the Nansemond, the Dan, and the Clinch. Grade school teachers, scientists in many disciplines, and professors at colleges large and small have taught their students about the value of protecting and conserving the natural world, and many have conducted policy-changing scientific research to support advances in managing fisheries and other natural resources such as water quality.

The leaders of several larger environmental groups throughout Virginia have formed a statewide advocacy organization whose purpose is to rally its member groups to persuade members of state government to pass laws and appropriate funds to carry out the goals of Article XI of the constitution. Professionally staffed conservation organizations such as The Nature Conservancy, The Conservation Fund, and the Piedmont Environmental Council have all contributed tangible results to the conservation of land throughout Virginia. The Chesapeake Bay Foundation—and its thousands of members—has focused on cleaning up and preventing future pollution of the marvelous natural resources of the Chesapeake Bay. It has documented the condition of the Bay and championed its cleanup for decades.

THE ENDOWMENT set out to make a difference, not just a contribution. Its "first dollar" approach to funding helped to start many of the local and statewide groups that have become so effective today. The Endowment realized that if a first grant to an organization went well, then additional support would follow. Much of that support is due to the VEE board's ability to focus its grant-making on a few key areas and "give 'til it helps." Typically, we and other foundations tended to make grants of one year's duration. Pretty quickly we discovered that some projects either needed more time to produce results or in other cases were doing great work right away, and we wanted to encourage them with more support. We did not set out to make multiyear grants to groups, but that is often what happened. We were flexible, focused, and results-oriented. Leading grant-making foundations from across Virginia and the region have contributed many millions of dollars to support these various efforts. Fifty years ago, there was not a single foundation devoted entirely to environmental work. Now, many are, and many more have added the environment to their priorities for grant-making. Yet, despite all the progress made so far, pollution remains a threat to health and the environment.

After fifty years of new laws, policies, and programs as well as other positive accomplishments, environmental quality in Virginia is in good condition and getting better. The history of how that all happened is an engaging tale full of heroes and heroines persisting over a long period of time to bring us to this partial state of environmental grace. This story is ultimately about them and what they have accomplished.

They are people who saw an environmental problem and decided to do something about it. Few of them were environmental experts when they started. Mostly, they had the determination and persistence over many years to solve environmental problems, change the laws, and challenge the regulations, all in the service of preventing pollution, cleaning it up, and conserving their part of the natural environment—to preserve their sense of place. This book is about them, and this book is for them.

The ripple effects of Judge Merhige's original idea continue to benefit the people of the Commonwealth. I am not a historian, and my narrative of these events is neither completely objective nor comprehensive, but I recognize the need for a record of these events, the great work that many people have accomplished over the past five decades.

The Endowment's work continues to this day. By focusing and leveraging its funds, this small foundation continues to have a significant impact in helping to build the environmental movement in Virginia, empowering citizens to assert their constitutional right to a clean environment. Think of this book as an example of how to make environmental progress at the state level—a model for what works, complete with specific stories, lessons learned, and mistakes made.

The importance of the Kepone disaster is that it uncovered the problem of undocumented toxic wastes being discharged into the air and water. The significance of Judge Merhige's decision in the case is what happened next, the rest of the story. If you see a lesson to be learned or an example to be followed, your actions will further leverage the idea that good can come out of bad, that there is opportunity in adversity, and that the disaster does not have to be the end of the story.

# 1

# A Poisoned River

## THE DISASTER THAT STARTED IT ALL

ANALYSIS OF a single human blood sample sent to the Centers for Disease Control and Prevention in mid-1975 led to the discovery of the Commonwealth of Virginia's biggest environmental disaster. The James River was poisoned with Kepone (chlordecone), a persistent chlorinated hydrocarbon used to kill agricultural pests, manufactured in Hopewell, Virginia. Kepone was being discharged directly into the river by a small company known as Life Sciences Products Company (LSP), and before that, from 1966 to 1974, by Allied Chemical Corporation. These discharges violated the federal Clean Water Act and extensively damaged both the environment and the local economy. LSP, established in 1974, was a spin-off of Kepone's previous manufacturer, Allied Chemical Corporation. It was run by two former employees of Allied Chemical, and Allied was its only customer.[1] The US government sued both companies.

Kepone is a white powdery substance that workers handled and were otherwise exposed to throughout their shifts, breathing it, their clothes covered by it. Many of them developed various nervous system illnesses, which they called the "Kepone shakes." One day, a doctor in Hopewell sent one of the workers to see a cardiologist, who suspected the worker might be suffering from chemical poisoning and sent samples of the man's blood and urine to the Centers for Disease Control (CDC) in Atlanta for analysis.

The lab analysis of the sample by the CDC showed high levels of Kepone in the worker's blood. The CDC notified the state epidemiologist, Dr. Robert Jackson of the State Health Department, of these results. Dr. Jackson visited Life Sciences Products, examined some of the workers, and concluded that the State Health Department needed to act. The plant was ordered closed on July 24, 1975.

LSP, and Allied before it, had been discharging millions of pounds of Kepone-laden wastes into the nearby James River and its tributaries, so in addition to the risk to human health there was an environmental emergency as well. In December 1975, Virginia Governor Mills Godwin banned commercial fishing in the James from Richmond east to the Chesapeake Bay, a distance of seventy-eight miles.

In 1976, as a result of the Kepone disaster, Virginia enacted its Toxic Substances Information Act,[2] which required companies manufacturing and using chemicals to disclose the details to the state government, while mandating that the state keep such information confidential and not publish it. This was quickly followed at the federal level later that year by a more comprehensive Toxic Substances Control Act (TOSCA), a milestone that occurred fourteen years after the publication of Rachel Carson's *Silent Spring,* the book that first brought worldwide attention to the human-made pollution that was imperiling the earth.

Meanwhile, litigation in federal and state courts proceeded against both Allied and Life Sciences Products. The federal government's case against Allied Chemical stated that the company "unlawfully discharged and deposited from its alleged Hopewell establishment industrial wastes from the production of octahydro1, 2, 4 metheno2Hcyclobuta [cd] pentalen2 one [hereinafter Kepone] into Gravelly Run, from which the refuse was washed into the James River, a navigable waterway of the United States."[3]

Pollution remains a threat to health and the environment, because state and federal environmental agencies continue to issue permits to discharge toxic waste material into the air, land, and water of the United States.

The damage caused by Kepone was the largest environmental disaster Virginia had ever experienced. Economic damage to the commercial fishing industry was severe. Even now, well into the twenty-first century, Kepone remains at measurable levels in James River sediments and is found in trace amounts in fish. The State Health Department still has fish advisories on the James for PCBs (polychlorinated biphenyls), mercury, and Kepone. Kepone continues to be manufactured overseas, and its use on banana plants in the French West Indies has given rise to serious health concerns there. The United Nations Environmental Programme has called for an end to its use.

# 2

# The Judge's Trust

HOW AN UNPRECEDENTED COURT RULING CREATED
THE VIRGINIA ENVIRONMENTAL ENDOWMENT

THE FEDERAL CASE against Allied Chemical Corporation was placed on Judge Robert Merhige's docket in the summer of 1976. Judge Merhige had an outstanding reputation and was known for resolving complex and controversial cases. His judicial career was marked by epic engagements with the most intense controversies and litigation of his era, such as school desegregation in the City of Richmond, protests at Wounded Knee, the 1972 Watergate scandal, and ordering the University of Virginia to admit women.[1]

Robert Reynold Merhige Jr. was born in Brooklyn, New York, of Lebanese and Irish ancestry. Despite being well under six feet tall, his talent on the basketball court got him into High Point College in North Carolina on a basketball scholarship. After college, he graduated from the University of Richmond law school in 1942. He served in the Army Air Forces in World War II, returning with an Air Medal with four oak-leaf clusters. After establishing himself in Richmond, he became one of Virginia's most successful criminal defense lawyers. He would later also receive a master of laws degree from the University of Virginia in 1982.

Appointed by President Johnson in 1967, Judge Merhige served on the US District Court for the Eastern District of Virginia. His magnificent courtroom was in the Italianate-style antebellum federal court building on Main Street in downtown Richmond. He took senior status in 1986 but continued hearing cases for many more years. He loved the law. He claimed he hadn't worked a day in his life since he entered law school—the law was his passion, not work.

IN 1976 Judge Merhige was well known for his 1972 decision to order school desegregation. The personal price he paid in the aftermath of that decision made him an icon of judicial courage and independence. He was given twenty-four-hour protection by federal marshals, as threats

Judge Robert R. Merhige Jr. (Photograph by Kent Eanes; reproduced by permission)

of violence were frequently made to him and his family, extending as far as the shooting of his dog and the burning of a guest cottage on his property.

Cases in the eastern district were assigned on a rotating basis among the several district judges. Talking with him years later, I asked how it happened that the most difficult cases often turned up on his docket. "Just lucky, I guess, I'm the one whose name keeps coming up," he said as he smiled, winked, and changed the subject.

Over the summer of 1976, the Kepone case had become a big pollution story and was thoroughly covered in the national, state, and local media. Some weeks it seemed that CBS News was reporting on it nightly, with Walter Cronkite, its weeknight anchor, often pronouncing "Kepone" similarly to the name of the famous 1930s-era Chicago gangster Al Capone. I suspect that TV producers were not in a position to tell "the most trusted man in America" how to pronounce words.

The trial came to a temporary halt on August 19, 1976, when Allied Chemical Corporation pleaded no contest to 940 criminal counts of discharging wastes from the production of Kepone into the James River. Allied had previously argued that Life Sciences Products, a spin-off from Allied, was principally responsible for the Kepone pollution of the river, but evidence from the fish samples routinely collected each year by the Virginia Institute of Marine Science demonstrated that Kepone was accumulating in James River fish during the time Allied manufactured the substance.

Judge Merhige fined Allied $13.24 million for the illegal dumping. At the time, it was the largest water pollution fine ever levied. The US Attorney who prosecuted the case, William Cummings, recalled that moment: "It was a major event when Judge Merhige announced the fine would be the maximum allowed on all the 940 counts that Allied had pled nolo contendere. It was a shock, I think, to everybody and brought a real awareness, because we saw magazine articles in business magazines thereafter that other companies were now finally taking notice. They were saying people can get in trouble if you don't pay attention to the regulations. That's the message he wanted to go out and I think it did go out."[2]

Judge Merhige also expressed a desire that the parties to the case—Allied, its lawyers, and the US Attorney—discuss some ways that the outcome of this case could include benefits for Virginia, where the damage took place. He noted that all of the money from payment of the fine was destined for the federal treasury, taking issue with how this left no money for the people of Virginia.

On February 1, 1977, the parties reconvened in Judge Merhige's courtroom to present an offer that would benefit Virginians. After months of negotiations among both the government's and Allied's lawyers, an unprecedented idea was offered to the court. The company volunteered to put up $8 million to endow a new nonprofit corporation under section 501(c)(4) of the federal tax code. Allied's first proposal was to set up a trust for the environment, one they would control. Judge Merhige rejected the latter part of the proposal and instead required the creation of the independent entity that would be known as the Virginia Environmental Endowment. It is still powerful to recall how novel this idea was. Judge Merhige's conclusion to this case was unprecedented.

The Endowment's purpose would be to use the money to improve the quality of Virginia's environment.[3] Judge Merhige thanked Allied for the idea and promptly accepted the proposal, causing the creation of the Virginia Environmental Endowment. At this point, Allied was obligated by fine and by the voluntary contribution to pay $21.24 million, but even though he did not have to, and in response to Allied's gesture, Judge Merhige reduced Allied's federal fine by $8 million. Allied still paid out $13.24 million, but $8 million went to fund VEE, so that only $5.24 million was paid to the federal treasury, where, at least in Virginia, nothing further was ever heard about it.

The Virginia Environmental Endowment, on the other hand, used that initial $8 million during the next four decades to spend approximately $29 million on more than 1,400 grants for environmental projects. When supplemented by the matching funds VEE has always required for its grants, the total added up to about $70 million in solid environmental accomplishments—and when I retired in 2013, the endowment still had over $19 million in the bank to continue its good work.

Judge Merhige's fine was the largest ever imposed for violating federal environmental laws. The judge stated, "I did so because that was the maximum fine I could levy." Several weeks after the decision, when I first met him, Judge Merhige told me that this pollution was the worst he had ever seen. As long as he had anything to say about it, he continued, he would never allow this to be repeated.

Inside the courtroom, an effort was underway to clarify the intention of the fund. As part of the "legislative history" of this new corporation—that is, the court proceedings of February 1, 1977—the lawyers all agreed that the funds would not be limited to Kepone matters. The drafting and debates of bills in legislatures constitute the legislative history of a law. In this case, the discussion in Judge Merhige's courtroom constituted a comparable "legislative history" for how and why and with what dimensions the Virginia Environmental Endowment was created. William Cummings, who had initially opposed the idea of the Virginia Environmental Endowment, went further, stating to the court, "We are very concerned that the money spent by this fund is not used to alleviate or reduce Allied's civil liabilities as they may be established." Judge Merhige replied, "That is one of the reasons I asked you to serve on the board, Mr. Cummings, to be alert to that." The judge went on to clarify the purposes of the new fund, which had been mischaracterized as a "Kepone fund" in some news accounts: "It is not intended to be used in any manner to reduce the legal liability of Allied; one. Two, it is not limited to alleviating the effects of kepone. There are other factors."

Judge Merhige was not only VEE's "founding father," but he was largely responsible for appointing its board of directors. Speaking on behalf of the defendant, attorney Murray Janus stated, "The foundation of this endowment, and the reason it is going to work, if your Honor please, is because of the confidence that we have in his Honor in appointing those types of directors that are going to have the integrity and the independence to make it work." The court then spoke to the independence of the board: "My control will dissipate as soon as I have appointed the board. This board is not to report to me. The members of the board understand that. They are independent." Judge Merhige did a superb job of selecting those who would guide the Endowment. Speaking years later at a University of Richmond seminar about the Kepone case, Judge Merhige recalled:

> In any event, I asked the U. S. Attorney who prosecuted the case, Mr. Cummings, if he would get on the board because under the trust the Senior Judge in the Richmond Division appoints the board of trustees. I asked Mr. Cummings to get on because I knew he was thoroughly familiar with the case, and I didn't want any of the funds used to help Allied buy off their civil liabilities. He accepted. As I recall, I appointed Judge Henry MacKenzie, who was an avid sportsman and very much interested in our environment; Admiral Ross P. Bullard, who was the Coast Guard Admiral in charge of the navigable waters around the Chesapeake Bay and the James River, so he was thoroughly familiar. Then I was fortunate enough to get Sydney and Frances Lewis, whose names may be familiar to you, who knew how to spend money from what I read in the paper. Cathy Douglas, a young lawyer who was the wife of former Supreme Court justice William O. Douglas. Then finally a banker. I thought we needed a banker. George Yowell, president of Dominion National bank, accepted. They were a great board.[4]

The criteria Judge Merhige used for selection for board members were never written down, but he is known to have said many times that they did not have to be environmental experts and that they must come from diverse backgrounds, have wide knowledge inventories and experiences, possess good judgment, be independent, and be committed to the mission of improving the environment. The Endowment executive director's job was to educate them about the environment and to draw upon their knowledge and experiences to do the right things and do them well.

Judge Merhige was adamant that VEE be an independent entity, beholden to no one. He emphasized VEE's independence both from Allied and from any government, federal or state. In fact, the bylaws stated that no one associated with Allied could ever serve on the board of directors of the Endowment. He stated on the record that he would not have accepted

Allied's generous proposal "if I believed one penny of the fund was going to be used to reduce legal liability on the part of Allied." Independence was a principal criterion on multiple levels: "I know individually they are all independent. I have known all of them long enough to know that. But collectively they are going to be independent. They have no obligation to the court. I am trying to be careful by explaining that it was not the purpose of the Court in reducing the fine so that Allied could use this fund to reduce any of their legal liability. They understand that. . . . What the (board members) do after they are appointed is strictly up to them. My control will dissipate as soon as I have appointed the board. This board is not to report to me."

VEE has been fortunate since its beginning to attract more than two dozen outstanding board members committed to its mission and dedicated to its implementation. The judge appointed William Cummings, the prosecutor, as VEE's first chairman—specifically to make sure the foundation focused on pollution prevention and environmental improvement and not to alleviate Allied's responsibilities for remediation in any way. Over the years, Judge Merhige appointed many outstanding trustees who have served the Endowment well, including former governor Linwood Holton, former first ladies Jinks Holton and Jeannie Baliles, Paul U. Elbling, Dixon M. Butler, Alson H. Smith Jr., Patricia Kluge, Byron Yost, Robert Smith, Nina Randolph, Robin Baliles, Robert M. Freeman, Cathleen Douglas, and Tom Wolfe, a Richmond native, author, and astute chronicler of American society. Jinks Holton succeeded Bill Cummings as board chair, followed by Dixon Butler.

The Endowment's story reflects the board's excellent judgment and ability to see where the Endowment could really make a difference, as well as its remarkable level of engagement. Most ideas or recommendations I proposed to the board were made better by their discussion of them. VEE's current board members continue this strong tradition of leadership.

JUDGE MERHIGE often claimed that he was not sure who first thought up the Endowment but that all the parties deserved credit for it. The creation of the Virginia Environmental Endowment was itself an original idea. I think the Kepone case bothered Judge Merhige but also was a revelation to him, enabling him to see how pollution of the environment hurt people as well as fish and wildlife and that it hurt the economy with its high costs of remediation and the closing of the James River to commercial fishing.

He expressed his dissatisfaction with the pollution that had occurred and made clear his desire to never see such a thing again. He believed there was a role for the federal courts to play in preventing pollution by holding polluters accountable for their actions. He found a way to fine the polluter and do some good for Virginia too, an unprecedented outcome. We know now how things turned out, but we didn't know then how the effects and benefits of this idea would spread throughout the Commonwealth and beyond.

Judge Merhige retired from the bench in 1998 and joined the law firm of Hunton & Williams in Richmond. Former Virginia governor Gerald "Jerry" Baliles had the office next to the judge. They were great friends, as illustrated by a story Jerry Baliles used to enjoy telling: "We pulled up in my low-numbered license plate Cadillac with lawyers and judges standing there on a summer-like day. The judge rolled down the window, invited his friends to get in the car and asked with a mischievous grin, 'Have you met my chauffeur, the former governor?'"

One time, I went to have lunch with Judge Merhige in his beautiful chambers in an 1850s Italian Renaissance–style federal building. We knew each other quite well by this time. He had ordered sandwiches for us: "Hope you like ham and cheese." He sat behind his desk, and I sat across from it. I felt very comfortable chatting with this accomplished and respected man. We shared a New York City Irish American heritage and, as successful as he was in life, he never impressed me as someone who thought of himself as a "big deal." As we were finishing lunch, he looked at me and said with a slightly formal tone and with a big smile, "Jer, you can call me by my first name—Judge!"

Judge Merhige died in 2005. He was a courageous, courteous, wise, and kind human being who loved the law and his family and enjoyed good food, good drink, and good company. For me, the moment that best sums up this great man occurred when he stepped down from the federal bench. The occasion was the unveiling of his official portrait, the one that would hang in the courthouse after he retired. To celebrate this occasion, a gala reception was held at a little art gallery on Grove Avenue in the west end of Richmond. A considerable crowd of admirers, friends, family, and colleagues attended the reception, including myself.

The painting, by John Court, is a wonderful portrait and a fine tribute. The party was delightful, with delicious food and lots of champagne. The buzz was lively and loud, a loving celebration of the judge. People mingled, circulated, and chatted, including a strong contingent of his former clerks, with professional and personal friends catching up and new acquaintances made. After the good-natured speeches had concluded and the party was winding down to a low hum, I caught the judge's attention. We walked over to the painting and stood before it without speaking. After a couple of minutes, I said, "Well, judge, what do you think?" He looked at me and then at the painting, smiled, and, gesturing with his hands spread apart, proudly replied, "I think he captured the magic!"

# 3

# An Independent Board

ESTABLISHING THE VIRGINIA
ENVIRONMENTAL ENDOWMENT

JUDGE MERHIGE enjoyed choosing board members. He had an instinctive understanding of who would do well in the role. His initial appointments to the board were all independent, experienced leaders. They ranged in age from thirties to sixties and included people in the retail business, a circuit court judge, a bank CEO, a retired coast guard admiral, the US Attorney who prosecuted the Allied case, and Cathleen H. Douglas, a young Washington attorney. The board's focused and results-oriented approach set in motion a record of accomplishment that owes much to their good judgment. They had the discipline to ask the right questions and the capacity to learn as they grew into their responsibilities. Over and over Judge Merhige stressed the board's independence—from Allied, from the state, from the federal government, from serving on boards of potential grantees, and from him.

The first order of business set the stage for all that happened subsequently. Board Chair Bill Cummings informed the Attorney General of Virginia that the Endowment's money could not be used for state remediation efforts: "We will follow the judge's advice . . . because the government of Virginia has a responsibility to its own citizens to use its own resources. This money was designed to help the citizens of Virginia, in a way that the government could not . . . We decided that was going to be our guiding principle, to use the money in a way that could not be used by government and tell the state that they had to clean up the James with their own resources."[1] The rest of the VEE story would not have happened if the board had decided otherwise.

When the Endowment was established, I had been serving as chairman and administrator of Virginia's Council on the Environment, a state agency responsible for coordinating environmental policies. I was in the middle of my eighth year, serving first Governor Holton and then Governor Godwin as head of the Council. The function of the Council and its membership had changed during the Godwin years. A Secretary of Commerce and Resources now oversaw and coordinated policy among both development and environmental agencies. The Council's role had been downsized from its role under Governor Holton, when it had been the lead agency for developing and coordinating environmental policy. Now, the Council functioned mostly to publish an annual report on the state of Virginia's environment and coordinate the review of environmental impact reports from state agencies proposing to construct new capital projects.

When he reappointed me, Governor Godwin had encouraged me to reach out to him directly any time I thought it was necessary, but the emphasis in his administration was clearly on the commerce side of the secretariat, and that's what he paid attention to. By 1977, with only a few months left in his term, I was looking at other possibilities.

Of course, I was aware that the Virginia Environmental Endowment had been established, but I had not heard much about it. That's when Louise Burke, spokesperson and lobbyist for the Conservation Council of Virginia, came to see me. She was a gracious middle-aged woman who toiled ceaselessly to persuade governors and legislators of the merits of protecting the environment. Her passion for water quality and a clean environment, as well as the set of skills she honed convincing the City of Richmond to protect the southern shoreline of the James River, led to the eventual establishment of the James River Park.

She wasted no time getting to the point of her visit: "Jerry, you need to apply for the job of executive director of this new environmental fund Judge Merhige set up. It is a wonderful opportunity to improve Virginia's environment, and I can't think of anyone better qualified to tackle that job than you. You need to apply. I'll write a recommendation for you."

One of the things I loved about Louise was how she could say the most consequential things in the nicest, low-key way. As a mild-mannered woman with silver-gray hair and big eyeglasses, people often underestimated her, but she was made of steel. I was flattered that she thought so well of me.

"I hadn't really thought about it," I told her, and I meant it. I had no idea what this new group was going to do or even who was involved with it. Louise said, "I will send in a letter telling this new board to interview you and give them plenty of reasons why they should do so."

Honest to God, that's how it started. Not long thereafter, I got a phone call from the US Attorney's office in Richmond. William Cummings was the US Attorney for the eastern district of Virginia, and he had also become chair of the VEE board. He said that he was holding a "wonderful letter of recommendation from Louise Burke. Would you be willing to meet with the Endowment's board of directors and give us some advice about environmental issues in Virginia?" I was happy to do so.

A couple of weeks later, we all met in a conference room in the US Attorney's office in the Italianate-style antebellum federal courthouse on Main Street. The building now houses the Fourth Circuit Court of Appeals, but in 1977 the District Court and its several courtrooms, the Appeals Court, the US Attorney's Office, and a Post Office branch were all packed into the building. I had never met any of the board members, although I had heard of two of them, Sydney and Frances Lewis, Richmonders who had founded a successful business called Best Products, Inc.

I entered the conference room and shook hands with the six members who were seated around a large rectangular table. All of us wore business attire. It was obvious right away that the board was a friendly group, as each member welcomed me warmly. I don't know exactly what Louise Burke had told them, but they made it clear they were eager to discuss possible directions for the Endowment's work.

One after another they asked questions and I tried to answer as best I could. Soon it turned into a discussion involving everyone in the room. For example, I was asked, what did I see as the most pressing environmental issue in Virginia? Who is supposed to do something about it? What are they doing? What else could or should be done? What role might VEE play in this, given its mission as a private nonprofit to improve the quality of Virginia's environment?

By the nature of their questions and their skepticism about how Virginia had handled the Kepone disaster, it quickly became clear that they wanted to honor the unique opportunity the judge had entrusted to them. They wanted to do things that would make a difference.

What did I think they should do? they wanted to know. We spent almost three hours talking, raising and answering questions. I cited examples of environmental matters in Virginia that I thought needed attention. I outlined a strategic management approach: "survey, decide what to do, assign responsibility, follow through, evaluate, and repeat." That approach captured their interest, notably—for a group of leaders who had little experience with environmental matters—the first step, to gather information from a variety of sources. We had a freewheeling discussion of the possibilities regarding what kind of information, where to get it, what to do with it, and how it might help the Endowment begin its work.

Finally noticing the lateness of the hour, Bill Cummings thanked me for my time and ideas and expressed the hope that we might meet again. I felt that these people could figure out something special and it would be fun to be a part of it.

About a week later, Bill Cummings called to offer me the opportunity to become executive director of the Virginia Environmental Endowment. I went to work on May 31, 1977, operating at first out of the US Attorney's Office, for the few weeks it took for us to find a small office space in downtown Richmond. At that point I hired Betty Toler, who quickly became indispensable. None of us had any foundation experience.

One question loomed paramount: "How could we make the best use of the money we have been given?" We studied the foundation world to see if there were role models for us to emulate and discovered that we were the only foundation in the country focusing 100 percent of our grants on environmental work. Other foundations were spending some of their grants on the environment, but we were the only one to focus on it exclusively.

We were unique in that regard from our inception. When we looked around to see who was funding environmental work, we also learned that only a few foundations were doing environmental policy work at all. Many foundations didn't want to touch it because it involved public policies about which people disagreed, and very few foundations at that time engaged in public policy issues of any kind. This was close to fifty years ago, and today environmental grant-making has increased substantially, as has policy-oriented work. For example, there's a group called the Environmental Grantmakers Association (EGA), which VEE helped create in 1987, ten years after we started. EGA's mission these days is to promote just, effective philanthropy for people and the planet.[2]

The board began by asking, "What needs doing?," already thinking about how a relatively small foundation might best leverage its limited funds to make a difference for Virginia. Without a comparable role model to follow, the board set its own distinctive course and had the humility to seek help in doing so.

During the summer of 1977, the board sought knowledgeable advice to educate itself about what we might do to be helpful. We met with scientists, policy makers, advocates, academics, and executive- and legislative-branch officials. We were particularly interested in specific suggestions and recommendations. People were generous in helping us to see the range of possible actions we could take. We quickly realized the impossibility of attempting it all and decided to focus on a limited number of issues where VEE might be able to make a difference.

As we felt our way into this new responsibility, among the possible approaches that we considered were the following: seeing if there were two or three significant problems we could address that would spend the money in a few short bursts for maximum short-term impact; hiring more expert staff and developing our own operating program for improving the environment of Virginia; or proceeding to operate similarly to a private, independent grant-making foundation, making a series of grants on priority subjects and topics over the next two or three years and then evaluating what, if any, difference we were making.

At this early stage, we were all just beginning to learn how to define what our work might be and had no clear consensus on how long we might even stay in operation. It was a great opportunity, but we didn't know whether we could make this idea of Judge Merhige's live up to its potential. Nor could we know that the decisions we were making about priorities and operations would turn out as well as they eventually did. We charted our own direction as an independent advocate for improving Virginia's environment.

On August 19, 1977, the board met at the Virginia Beach home of Sydney and Frances Lewis for a retreat to review what we had learned, sort through the many ideas we had heard, and make some initial decisions about how to get on with our work. We had complete discretion and flexibility to choose how to move forward. We and our spouses were treated to the Lewises' famous hospitality at their beachfront home, which was filled with modern art and sculpture, Tiffany lamps, and comfortable furniture.[3] Such a congenial atmosphere encouraged us all to relax and think big and creatively.

Not all the board members knew each other well at that point. Admiral Bullard and Judge MacKenzie, both from Portsmouth, were acquainted, but before being appointed to the Endowment board, neither of them had known the Lewises or the other board members. The Lewises knew George Yowell, but no one knew their new chairman, US Attorney Bill Cummings. The one person they all had in common was Judge Merhige. A weekend gathering in a comfortable beachfront home was just the ticket to forge collegiality and friendships that would continue for years to come. Ever after, the board made having a good meal together a fixture of all their board meetings, making them much more productive and enjoyable.

With $8 million to distribute (about $38 million in today's value), this was a considerable opportunity Judge Merhige had given us. The conversation flowed at breakfast, lunch, and dinner and throughout the day and evening. Many ideas percolated, and constructive disagreements refined them. Should we look for a couple of big projects that might have an immediate impact? Or perhaps take it more slowly as we learned our way around what needed doing and where VEE might be able to make a difference?

We had to consider just how long we wanted to be in business—a year or two, or indefinitely. The choice was up to us, and for the time being we chose to invest the money in short-term financial instruments like CDs, overnight paper, and thirty-day bonds and use the interest collected for grants. If you had told any of us then that the Endowment would still be making grants well into the twenty-first century, I doubt we would have believed it. It took us a while to see the potential of the opportunity we had been given.

The many ideas and suggestions we had heard in previous weeks needed to be narrowed down, because we couldn't do them all, not even by giving the entire $8 million away at once. We asked ourselves: Is science the most important factor? What about why and how the Kepone mess happened in the first place? Are the laws and regulations sufficient for the job of protecting the environment? How about the public, what do they need to know, and how do they find out what is being discharged into the environment?

All these relaxed, extended conversations fueled by delicious food prepared by the Lewises' chef—pizza from an outdoor woodburning oven at lunch; steaks at dinner—helped to define VEE's role and priorities, sort out what needed doing that we could help with, and clarify how to proceed in positive, constructive ways. The remarkable American and French wines from their collection were terrific for sustaining conversations too.

At the conclusion of the retreat, several choices became clear. Our mission would focus on preventing another disaster from happening. Our reasons for being were prevention of pollution and improvement, not remediation, of the environment. We decided firmly that VEE would not be a Kepone fund, in contrast to the suggestions made by some news accounts earlier in the year when the Endowment was created. No funds would be used to repair the damage that had been done by Kepone. That was Allied Chemical Corporation's responsibility.

The board adopted the following initial mission statement: "The Virginia Environmental Endowment (VEE) is a nonprofit, independent corporation committed to the improvement of Virginia's environment. The endowment hopes to become a catalyst by using its resources to help citizens, industry, and government take constructive action to enhance Virginia's environment." This statement signaled the endowment's willingness to work with organizations of all kinds that shared its interest in improving Virginia's environment.

We discussed how, because of my experience as a top state environmental official working for the previous two governors, my knowledge could be used on behalf of VEE to advise government officials as well as grantees and applicants. We could be "more than money" to help improve the environment. Over the years, I eventually served on many state boards and commissions, including the Chesapeake Bay Program Citizen Advisory Council (chair), the Virginia Conservation and Recreation Foundation (vice-chair), the Virginia Uranium Study, the Natural Resources Commission, and the Commonwealth Transportation Board (secretary), among others.

We chose three initial priorities: the effects of toxic substances on human health and the environment, environmental law and policy, and ecological research. As a fan of the first "environmental" president, Teddy Roosevelt, I was reminded of his advice, "Do what you can, with what you have, where you are." "Toxic substances and water quality" was an appropriate focus given the origin of VEE. At that time, the Clean Water Act was only five years old, and people were still trying to figure out how to implement its ambitious goal of eliminating discharges to rivers and streams, so our emphasis on this priority was ahead of the curve. Our decision to focus on utilizing environmental law was also a consequence of the judicial action that created the Endowment. And ecological research is always an important environmental need, one that requires more attention than it was receiving at the time. We did not view these priorities as restrictions. If some outstanding project came along that didn't fall into one of the priorities, we wanted to remain flexible enough to be responsive to it.

Another decision made by the board was to leverage its grant dollars. The board established a one-to-one matching requirement for grants, to encourage other support and to double the value of its grants. What is more, over time the "seal of approval" implied in receiving a grant from VEE has been a great help to many grantees seeking additional funds for projects.

We also deliberately focused on public policy, something most foundations avoided in those days. In the public policy arena, change takes time, and for environmental work, that can mean decades. However, operating in the public policy realm is where leverage can be maximized to change the laws—that is, to change the rules. We were interested in improving laws and public policies that would strengthen environmental protection. Given that Virginia had recently adopted a new constitution, containing Article XI, which states that "it shall be the Commonwealth's policy to protect its atmosphere, lands and waters from pollution, impairment or destruction for the benefit and general welfare of the people of the Commonwealth," it made sense for us to help figure out how that mandate might be implemented in general, and more specifically how our grants could enable people to help realize that goal.

At the conclusion of the beach retreat, the board chose to organize and operate like a private foundation, at least, as Bill Cummings put it, "until we learn more about what's needed and what's possible." This was an important choice for the board; it signaled that the Endowment was going to devote enough time and thought to address how best to use the funds entrusted to it.

The following month we asked the Council on Foundations, philanthropy's national advocate, to help the board organize VEE as a grant-making foundation. Eugene "Struck" Struckhoff, vice president of the Council and later its president, came to Richmond and spent a day teaching us how private and community foundations operate. He had written the most current book on the subject and was an excellent teacher, emphasizing the value of focusing on a few priority topics, having guidelines to follow, publicizing to potential applicants what we were hoping to fund, and always remaining professional, transparent, and respectful in our dealings with applicants. He helped us to understand the various roles foundations could pursue, beyond making contributions to favorite causes, in order to make a difference, have an impact.

He taught us to leverage our grant funds as much as possible, to make them go further and work harder, and he advised us to monitor grants periodically to insure good results. He noted that, while the federal tax code governed the distribution of charitable foundations' assets, VEE could vary the annual spending amounts as it saw fit and not be bound by the 5 percent payout foundations had to make, because it was not a charitable foundation but a tax-exempt social welfare organization described in section 501(c)(4).

From the beginning, the board staked out a middle-of-the road position, offering to work with all sectors—business, government, citizens—to improve the environment. It wanted to make clear that, unlike many groups that have the word "environment" in their names, the Endowment is not an advocacy organization. Instead of being a lobbying group, it builds bridges of communication among all the different groups interested in Virginia's environmental quality. It takes a moderate, constructive approach to problem-solving.

As one way to demonstrate that approach, the board also chose to examine mediation and other alternative dispute-resolution possibilities as a way to bring people together in resolving environmental disputes. The board sincerely believed that people could accomplish more by working together than separately. Over the summer, board members had heard much about the need to protect the environment. They realized during the retreat weekend's discussions that all the people we heard from agreed that pollution was a serious problem. These same people, however, disagreed over what should be done about it, who should do it, and who would pay the costs. The conclusion the board reached was that VEE was in a unique position to work toward bringing these different views and approaches together to work on the shared goal of improving the quality of the environment. The board saw this as an opportunity to find a new approach, one whose purpose would be to resolve rather than litigate complex, multiparty environmental disputes.

To put this idea in the context of the 1970s, lawsuits allowed by the Clean Air Act and the Clean Water Act had become a frequent way to resolve environmental disputes. Both laws provided for citizen suits to enforce these federal laws. It was our position that, although necessary from time to time, lawsuits were not always the best answer for complex environmental disputes. The board decided that it would rather try to promote mediation than fund lawsuits, because, in the words of board member Judge Henry MacKenzie, "It's not our job to choose sides."

Later that year the board saw an opportunity to implement this goal by setting up a special fund to pay for the services of a professional mediator in environmental disputes. We opened an account with a bank in Hopewell and deposited $25,000, which would be equivalent to $120,000 in 2023. Although this decision illustrated in a tangible way the Endowment's middle-of-the-road approach to environmental issues, we ended up closing the fund due to lack of interest in utilizing it. We determined that we would try harder to encourage mediation, because we felt strongly that this was a good idea.

The board also decided that it would take a statewide perspective, not just a James River–centered one. It believed that its mission was to benefit all Virginians, and it recognized that toxic water pollution occurred throughout the Commonwealth.

It also resolved to be a first-dollar grant-maker by launching new investigations of science and public policy, seeding ideas that showed promise, and taking the lead in risking funds on new ideas. We quickly made clear that we were ready to stay with good ideas and people for years.

Operating as a foundation, the board had the independence to use its limited resources to protect, conserve, and improve the environment. Foundations enjoy special privileges in our system. In return for freedom from paying most taxes, foundations are encouraged to use their funds to improve society over the long term. Having this liberty to focus on the long view enables foundations to take risks that neither shareholders nor taxpayers would tolerate, and to take the time to see if a new idea, service, or program can work.

The board approached its mission by "investing in people, focusing on results." When confronted with a new idea or organization, often their response came down to "Let's try it, see what happens." Often, we would try making other kinds of grants at the same time, so that they reinforced each other. For example, we pursued different tactics on reducing toxic pollution, making grants for research, education, and law and public policy, all combining in a multifaceted approach to address the needs we had identified.

Without question, we were feeling our way along and learning as we went. Our judgment about making grants improved as we gained more experience, which also included grants given to endeavors that were more promise than performance. This is not intended as a criticism of those programs; it is the decision-makers who make the call, and we are not always right in our choices. But we were quick studies and learned as we moved ahead.

Admiral Bullard helped me a lot as we began our work. He would drive up to Richmond from Portsmouth to have lunch with me and to listen to how things were going. He was a skilled executive and a great listener. We would enjoy a sandwich in my tiny office and review what had been going on and what was on the horizon. And when he spoke, his voice had the soothing, deep, resonant timbre of a classical music station host. I enjoyed listening to him.

He was helpful during our first year, because none of us were sure yet about how we were doing. Admiral Bullard was a great one to bounce ideas off. "Tell me how it's going," he would say. I tended to be a talker, but I learned to listen when he talked. "Have you thought about taking any board members along to talk with grant applicants?" he once asked me. Honestly, I hadn't, but it was, and is, a good idea. We were just seeing the earliest progress reports from our first grants, so it was too soon to see if our grant decisions were leading to good results. Seeing and hearing from the grantee organization in person was a valuable experience for board members, and a great way to get a sense of how the first grants were working out. Not all board members joined me to meet with applicants, but those who did found out more by such visits than they would have from simply hearing my reports at board meetings.

To track the progress of grants, we immediately set up reporting requirements. We required grantees to file quarterly narrative and financial reports to document progress and to request adjustments to the work plan or budget if new information required that. In the professional philanthropy world this is called "expenditure responsibility." It became an important part of the strategic planning process we employed to help achieve results. It was focused, professional, and results-oriented. It was also great fun, because it allowed us to maintain a personal connection to the people involved in the work, without undermining their leadership.

The VEE board took an active approach: it identified needs, set priorities based on those needs, informed people who might apply for grants what the priorities are, and sought out and offered grants to people who could implement them. In seeking help, we looked to a lot of experts, including national environmental groups. We also asked for advice from other foundations with more experience than we had. Sarah Chasis, long-time leader of the Natural Resources Defense Council's water quality program, was willing to answer questions whenever I called on her. Ned Ames, trustee of the Cary Trust, gave us valuable advice about the needs of the Eastern Shore of Virginia, and Bill Bondurant of the Mary Reynolds Babcock Foundation in North Carolina, a former top state official in North Carolina, was both helpful and wise, reminding me that although what we as grant-makers do is crucial, what is truly important are the people we help. They are the ones who do the work and deserve the credit.

For the most part, our grant spending matched priorities closely. Occasionally, the board made exceptions when outstanding ideas emerged that were outside the list of current priorities. As Bill Bondurant put it in an annual report describing Babcock Foundation grants, "You can see by our choices what a 'Babcock grant' looks like. We also make exceptions."

Within a year of our start-up, Cathy Douglas resigned, and Judge Merhige needed to appoint someone to succeed her. The judge liked the idea of having a celebrity on the board, preferably Virginia-based or connected with our state in some fashion.

Judge Merhige, who was good friends with Sydney and Frances Lewis, called them up and said, "I need your help. I need to replace Cathy Douglas." They suggested Tom Wolfe, the American author and journalist. The Lewises and Tom Wolfe shared a Washington and Lee University background and had known each other for decades. Judge Merhige thought it was a brilliant idea, because it met all his criteria—Virginia connection; wise, knowledgeable, and experienced; and independent.

Tom Wolfe was originally from Richmond, was an accomplished practitioner of the "new journalism," and was a celebrity. Judge Merhige thought the board members would enjoy working with him. None of the board members were environmental experts, but they each were superlative in their given fields. The judge wanted to promote collegiality and make it fun to serve on the VEE board. The board members were all busy people who were much in demand, and they served on the board partially out of love and respect for Judge Merhige.

In person Tom Wolfe was among the most charming, intelligent, and succinct people I've ever been privileged to meet. He could convey in one word what most of us need a paragraph to express. And at one board meeting he complimented me on the suit I was wearing, asking where I had bought it. For me, this was the equivalent of Ted Williams, arguably the greatest hitter in baseball history, asking me for batting tips.

During that first year, we often met in the US Attorney's office in the federal courthouse in Richmond. We would meet in the conference room and have tuna sandwiches and iced tea, or something similar. When Judge Merhige heard about it, he called me into his chambers. He said to me, "What's this I hear about sandwiches and iced tea down the hall?" I said, "Well, Judge, it's convenient to have our board meetings that way." He says, "You're not paying these people, right?" I said, "No, sir." "Well," he responded, "you need to compensate them in a way that they would appreciate. So, I'm telling you, when you have board meetings, build in relaxed time for letting them get to know each other. You need to give these people a good place to stay and a good meal. And that would be some compensation for all that they are giving you."

From that point on, we always did exactly that. And when Tom Wolfe was on the board, we met once a year in New York City for his convenience, and he always chose a nice restaurant. During one board meeting, we had dinner at a small restaurant in Manhattan, The Box Tree. It was a little jewel box of a restaurant: small, quiet, sumptuously decorated in deep forest-green walls and carpeting, with white tablecloths,

Virginia Environmental Endowment Board of Directors, 1980. *Seated, left to right:* Henry W. MacKenzie Jr., William B. Cummings, Frances A. Lewis. *Standing, left to right:* George L. Yowell, Sydney Lewis, Gerald P. McCarthy, Ross P. Bullard, Thomas K. Wolfe, Jr. (1980 VEE annual report)

professional servers, and delicious food and drink. In such an elegant setting—honestly, a little more upscale than our usual experiences of "good food, good wine, and good company"—the waiters began the cocktail hour by passing around flutes of champagne to whet our appetites. After a little while, I noticed that neither Judge MacKenzie nor Admiral Bullard had champagne, so I walked over to them, and before I could say a word, Judge MacKenzie asked with some intensity, "How do you get some bourbon around here?"

Operating independently of the government, VEE could pick its opportunities for where and how to carry out its mission to improve the environment. In terms of spending patterns, we preferred doing something about what was needed at the time rather than assuming VEE would operate in perpetuity. Instead of spending about 5 percent of assets each year as a strategy to preserve the option of perpetuity, we spent anywhere from 6 to 10 percent so that we would be able to put more money into current needs and projects. We also realized that improving the environment was going to take a long time and that a lot of people were going to have to work together to make lasting progress possible. Our middle-of-the-road approach preserved some future grant-making capacity without having to commit to it just yet, since we simply did not know enough at that time to make such a decision and there was no consensus among the board members regarding how long VEE might be in business. Perpetuity was put off in favor of the present.

Many years later, a future board decided to face the question of "perpetuity or not?" We considered what other foundations on both sides of that decision had done and examined the pros and cons of either choice.

A couple of foundations created and run by wealthy individuals had opted to close after a particular number of years in existence. While I had read some interesting analyses of the "perpetuity or not" question in the philanthropic press, I knew that in both cases, the decision was principally made because the donors wanted to see results in their lifetimes and have the money used for the subjects they were most interested in—because, ultimately, it is fun to practice giving while living.

Foundations where the family board was run by the third generation, or even later descendants, had already been institutionalized for the long run, and perpetuity was an apparent fact. "Independent foundations," those not tied to a founding family, had also at some point made a choice about perpetuity, either by consciously considering the matter or simply by managing their investments and grant-making with perpetuity the practical result.

At the time of this discussion, VEE had been in business for more than three decades, so we had a pretty good understanding of our business, our finances, and the needs that we could reasonably anticipate addressing. Our board chose to devote a special meeting to the subject of perpetuity. To prepare for that discussion, in addition to researching the professional literature on "perpetuity or not," I had talked with other foundation executives and family board members and prepared a paper summarizing the arguments for and against perpetuity.

As usual, the board members had read the material and were ready to discuss. Based on years of experience in seeing my ideas improved by their collegial contributions, I knew that I could expect one of the most substantive and delightful board discussions of our long tenure together. I summarized the arguments pro and con to start the discussion and then sat back and enjoyed the lively interaction among smart, thoughtful, fully prepared board members work its magic.

Bob Smith, the chair of the investment committee, approached the subject with finances as his focus: "What kind of investment strategy will we need if we opt for perpetuity? What happens to the money if we decide to terminate VEE?"

Each of the board members had long experience as volunteers with nonprofits in other fields besides environmental conservation; each had seen enough to know that VEE's decision whether to close or stay in business had implications well beyond our doors. For example, if we chose to spend down the assets over a short period of time, perhaps three to five years, we would rearrange our investment objectives and most likely make larger grants to fewer organizations. Nina Randolph and Robin Baliles wondered what the result would be if we transferred our assets to some of the groups we had supported for the longest time: "Is that the best course, or should we focus on a few environmental topics instead?" Even the thought of transferring all the money to a community foundation, to manage it going forward, was discussed.

After more than three hours of lively conversation, examining the pros and cons of various options, Anna Lawson, who had also previously served on conservation nonprofit boards, weighed in. It came down to this: "A grant from VEE is like a *Good Housekeeping* seal of approval for our grantees. It gives them credibility in seeking additional funds from other foundations. We could transfer our financial assets to other grant-making or grant-seeking organizations, but we can't transfer the good will and 'seal of approval' our grants impart to their recipients." The board unanimously chose perpetuity.

The importance of that decision became clear when we saw that if perpetuity was our corporate goal, then we needed a financial and investment plan aimed at supporting perpetuity. Previously, we had followed a "pragmatic perpetuity" approach whereby some years we spent more than 5 percent and some years less. Our investment objectives were optimistic too. The decision for perpetuity clarified the reality of what both our investment and spending policies had to be: lower return expectations and lower spending.

Finally, because none of us can serve in perpetuity, we realized that we needed to initiate succession planning. When I retired after a thirty-six-year tenure, all these pieces fit together nicely.

# 4

# Defining the New Endowment

NOT LONG AFTER we announced that we were ready to review proposals and asked people to send us their ideas, we received many proposals asking us to support various legal and public policy studies as well as scientific research on water pollution and toxic substances. This collection of ideas represented much more in requested funds than we could grant, a reality that continued in subsequent grant cycles. Gene Struckhoff at the Council of Foundations had warned us we would receive more requests than we could ever support, but as the Bretons say, you don't have to drink the whole sea.

How did VEE decide on a limited number of grants in the face of so many requests? We wanted to make choices based on our guidelines and criteria (as opposed to strict rules), but mostly on the priorities we had chosen.

It was not easy. We were still figuring out how to do things, and many of the proposals seemed worthy of supporting. Even though we had spent time establishing priorities, guidelines, and criteria, we knew that we didn't know much yet about what really needed doing, as evidenced by the variety of proposal topics we received in our first round of grant-making.

Several requests were for studying the effects of toxic substances on water quality. We also received proposals to study the effects of Kepone on fish and animals, even though Judge Merhige and the VEE board had made it clear that the Virginia Environmental Endowment was not a "Kepone fund."

I reviewed all the proposals and was prepared to present summaries of them to the board. I felt confident in talking about the proposals with the board; they had asked for my summary of each one, in place of reading the full proposal. That was a good choice, because we had not specified any maximum length for proposals and some of them ran to more than fifty pages. After that experience, we specified what we wanted in a proposal, including its maximum length.

This was the first meeting where the board would make decisions on which proposals would receive grants. By this time, the board members all knew, respected, and liked each other. None of us had governed a foundation before, but in my case, this review process was not very different from my experience in the Air Force in awarding research and development contracts. The board members entered easily into a lively discussion about the proposals. It was very conversational and informal. It was interesting to observe and react to the different styles and questions posed by the board. Bill Cummings, in his role as chair, mostly kept quiet except when someone wanted to know if a proposed grant might violate our charge to avoid anything that Allied might be responsible for. We also disposed of several proposals that were clearly Kepone remedial projects, for the same reason. Mrs. Lewis asked thoughtful questions about each proposal under consideration. Mr. Lewis asked, "Mr. McCarthy, can you tell me what will be different a year from now if we make this grant?" Admiral Bullard was impressed with this initial round of proposals and was curious about how some would work in practice. The board was thorough, and I felt we gave each proposal a fair chance. We discussed which ones would be a good fit for us and why, which ones were preventive rather than remedial, and we discussed where we might be able to make a difference.

The board made its first grants in December 1977. A grant to the University of Virginia Department of Environmental Sciences was used for a study at Lake Anna to address the origins of heavy metals in the lake's fishes and to determine the extent of heavy metal contamination in the sediments of Contrary Creek, which feeds into the lake. Another grant went to the College of William and Mary's Marshall-Wythe School of Law to conduct four environmental law conferences related to water quality. Virginia Tech researchers received a grant to develop natural means of disposing of toxic pesticides.

We tried to make good choices, but we didn't know, and wouldn't know for a while, whether those choices would work out well or not. Rather than certainty about our choices, we felt hopeful and confident about the people we were able to help.

The Endowment's flexibility was demonstrated in one of our first grants. It was a challenge grant to The Nature Conservancy (TNC) to establish a headquarters on the Eastern Shore of Virginia to manage and protect its multimillion-dollar investment in the offshore barrier islands and to establish its first community-based conservation program. These islands and the adjacent marshes constitute one of the largest coastal wilderness areas remaining on the East Coast of the United States. This grant was outside our initial priorities, but it was such a good idea—with excellent leadership, leverage, and partnership possibilities—that we decided to consider it.

The request was for $150,000 over a couple of years, $733,000 in to-day's dollars. The Conservancy would have to raise an equal amount from within Virginia to receive it. The Nature Conservancy had recently acquired ownership of most of the barrier islands off the Eastern Shore. As part of their stewardship, they wanted to make sure they had somebody on-site to oversee this program, develop good relations with the local community, and look after the islands.

Even though this proposal appeared to have little to do with our initial focus on toxic substances and water quality, it was a bold idea. The people involved were proven leaders, so we thought, "Well, it is a big grant, but if they can match it with another $150,000, then maybe we'll be able to do some good here." Also, we didn't pretend to have all the knowledge about environmental needs; there were plenty of other people out there who could bring us important ideas we weren't smart enough to think of. This was a good example of that.

President of The Nature Conservancy Pat Noonan and investment banker and conservationist Jim Wheat were capable and persuasive people. Noonan, a rare person who had both a master's degree in planning and another in business administration, had decided to spend his life conserving exceptional landscapes and natural habitats and protecting water quality. In a short time, he had built TNC into a confident and successful national conservation powerhouse. And to have a longtime Nature Conservancy supporter, Jim Wheat, as his partner in raising the other $150,000 for this project in Virginia seemed like a relatively low-risk grant, even as it was a large amount of money for us.

I had the pleasure of knowing both Noonan and Wheat for several years. Whenever I met with Jim Wheat, who was blind, he always jumped up at his desk greeting me with "Great to see you, Jerry, great to see you." Two of our board members, Sydney and Frances Lewis, also knew him well. We were confident that, together, Pat Noonan and Jim Wheat would successfully raise the money and get the work started. And, because this was a challenge grant, if they didn't reach the financial match, we didn't have to put up our funds.

As it turns out, Wheat underestimated; he didn't raise $150,000, he raised $300,000! That is the equivalent of about $1.36 million today.

TNC's Eastern Shore operation is one of the great conservation success stories. Now called the Virginia Coast Reserve, the barrier island complex is fully protected from development and conserved for the long run in Nature Conservancy ownership and federal agency ownership of the islands TNC does not own. The Virginia Coast Reserve is an international biosphere reserve and the longest expanse of coastal wilderness remaining on the East Coast. It is a model for how conservation can help a landscape to adapt and become more resilient in the face of a changing climate. It was designated a National Natural Landmark in 1979.

The Endowment, as a first-dollar funder at the start of programs that often remains to sustain outstanding programs over time, continued its support for the Eastern Shore Coast Reserve for many years. It also supported The Nature Conservancy's Virginia state chapter's work to create the state Natural Heritage Program and awarded it a $100,000 grant in 1995 to help launch the Clinch River Reserve Program in southwestern Virginia, both of which are national models.

VEE also awarded a third kind of grant, foreshadowing a major interest in the empowerment of conservation groups. In 1978 the Endowment made its first grant to the Conservation Council of Virginia, a statewide coalition of dozens of mostly small, local environmental volunteer groups. The Conservation Council's principal operating function was to coordinate legislative lobbying by its member groups to improve public policy on the environment in Virginia. VEE helped the Council for three years, which allowed it to hire a full-time executive director, publish a regular newsletter, and provide accurate information to its membership about legislative activities. It was VEE's hope that over time the capacity of the Council would grow if it developed strong board leadership. That took a while. Eleven years later, the Endowment was asked to help again, and that led to the creation of the Virginia Conservation Network in 1993, an institution that is now flourishing.

VEE's grants, though small compared to the needs, functioned like a trim tab on a boat or ship, a miniature rudder that augments the function of a boat's main rudder, providing leverage that affects the direction of the ship—in our case, focusing and leveraging the effects of small amounts of money in the public policy arena in such a way that their effects can be maximized, thereby creating the most change with the least number of resources.

Back then, there was no Chesapeake Bay Foundation in Virginia, no Southern Environmental Law Center, and no Friends of the Rappahannock, and even the newly formed Lower James River Association was strictly a volunteer organization. Except for the Virginia Outdoors Foundation, set up by the state in 1966, and the Piedmont Environmental Council, which was formed in 1972, there was not much land conservation taking place either.

We set out to do something about that, because a clean environment is a public good and a public right. Article XI of Virginia's constitution says as much: "It is the policy of the Commonwealth to protect its atmosphere, lands and waters from pollution, impairment or destruction." We believed that the public ought to have a voice in decisions about the environment. We seeded the establishment of many new environmental conservation organizations over the coming years, giving the public and the public interest a seat at the legislative and regulatory table for the first time. This is one of VEE's lasting legacies: the groups that we helped start continue to thrive today.

## The Institute for Environmental Negotiation

By late 1978, we wanted to see if there was a place for resolving environmental disputes constructively, rather than by the "win-lose" result of litigation. My investigation into the matter eventually led me to one of my alma maters, the University of Washington in Seattle.

Dr. Jerry Cormick was a research professor at UW and a former mediator for the US Department of Labor. He had plenty of experience using alternative dispute resolution (ADR) techniques to resolve complex labor problems. He had an insight that these same ADR skills could be applied to complex environmental and land use disputes in the Pacific Northwest. His efforts improved situations significantly and resolved several complex, multiparty disputes. Dr. Cormick pioneered the idea of transferring alternative dispute resolution skills and practices to environmental disputes. "Win-win" and ADR were new ideas at the time, and they appealed to VEE's sense of getting people to work together to improve the environment. Dr. Cormick was the only person doing this kind of work on environmental matters.

I called him up and invited him to come to Richmond and tell us about what he was doing. When I reviewed the example cases that he sent me before we met, I had a pretty good idea of the complexity of his cases and why it would take more than win-or-lose litigation to resolve them. He arrived a week or so later and was happy to help us explore the potential for environmental mediation as he practiced it out in the state of Washington.

Dr. Cormick described cases he was involved in, some of which involved a couple of dozen people representing different points of view gathering in the same room to hash things out. He insisted that they listen to each other as they would wish others to listen to them. He was the one who suggested that we look to a public university in Virginia as the home for an environmental mediation program, principally because of the need for an institutional base to support the idea.

A few weeks later, we had established what we hoped to accomplish and developed a way of trying out the idea. We issued a request for proposals, which we sent to universities that had expressed interest in the idea. We tested the idea with two universities, and in 1980, VEE partnered with the University of Virginia School of Architecture to establish the Institute for Environmental Negotiation (IEN). The Institute was the first such environmental mediation institute to be established at a university. The board decided on the University of Virginia in part because of the strong institutional support it offered. VEE offered to put up one-third of the cost of operating the new Institute, and UVA agreed to match that amount. The rest of the funds would have to be raised annually.

Dr. Richard Collins, a professor of planning in the Architecture School, became the first director of the organization. His wide knowledge and diverse experiences—not to mention his enthusiasm for the idea—made him a perfect choice to start the program. Before long, he and associate professor Bruce Dotson were involved in mediating several disputes, mostly centered on local land and zoning changes sought by some and opposed by others.

Dr. Collins had a gift for getting people into the same room to say what they thought and what they wanted. Eventually, he realized that it wasn't just the decision-making processes that often led to "win-lose" outcomes. To get to "win-win," he believed, the underlying public policies were the places to effect change. This insight led to the creation and practice of "policy dialogues" that involved many people and interests working their way to consensus about, for example, hazardous waste disposal, protection of the coastal shoreline to reduce water pollution, and water resource issues. The Surface Water Management Act and the Chesapeake Bay Preservation Act, both enacted in 1988, were successful results that emerged from this policy dialogue process.

After many years, Dr. Frank Dukes succeeded Rich Collins as director of the Institute and in turn was succeeded by longtime staff member and former associate director Tanya Denckla Cobb, who still occupies the role today. The decision to fund the establishment of the IEN was a tangible expression of our hope that the Endowment's middle-of-the-road approach to environmental issues would accomplish more than people suing each other. IEN is now one of the premier institutions of its kind in the country. In 2019 the Institute changed its name to the Institute for Engagement and Negotiation to reflect its wider scope of operations in several areas. The Institute was the signature achievement of this early period.

# 5

# Article XI

THE CONSERVATION ARTICLE

In 1971 Virginia voters overwhelmingly approved a new constitution for the Commonwealth of Virginia. The new constitution made explicit provision for protecting Virginia's air, land, water, and natural resources from pollution, impairment, or destruction, by including a conservation article, Article XI. Years later, two more sections were added, one relating to holding the natural oyster beds of the Commonwealth in trust for the people, and another concerning the right of the people to hunt, fish, and harvest game. It is the first two sections that concern us here.

I am grateful to Professor A. E. Dick Howard, principal author of the 1971 Virginia Constitution and the Warner-Booker Distinguished Professor of International Law at the University of Virginia. He served as executive director of the Commission on Constitutional Revision and directed the successful referendum campaign for the new constitution's ratification, which took effect in 1971. He graciously permitted me to excerpt information and analysis from his extensive explanation of the purpose of Article XI.[1]

Occasionally, one hears the argument that Article XI is aspirational rather than mandatory. Not long ago, I was talking to a friend, a distinguished lawyer, who read Article XI as "aspirational." I said, "If I said to you, if I was your boss and I said, you shall this take care of this matter, do you consider that aspirational or a mandate?" And he says, "Well, when you put it that way." And I replied, "Well, what other way is there to 'put' it?" And he says, "I see your point." I said, "I've discussed this with the gentleman who wrote it, Dick Howard, and he says that 'shall' means mandate, not an aspiration."

Professor Howard makes this case eloquently:

As a formal statement of the public policy of the Commonwealth, section 1 requires no implementing legislation. As public policy it became effective when the Constitution became operative on July 1, 1971. Section

1's self-executing quality is recognized by Section 2, which, in authorizing the Assembly to act, says that legislation is to be "in the furtherance of such policy" in existence by virtue of section 1. An enunciation of public policy, unlike a rule of conduct laid down by legislation, is not aimed at the private citizen and imposes no duty on him. Rather, it is a mandate for and a restraint on governmental activity. Section 1 is thus self-executing, not with regard to the public at large, but with regard to those entities which are constitutionally bound by public policy, namely, the government, its courts, and its agencies. Article XI is not, however, self-executing with respect to obligating the General Assembly to enact environmental legislation.

Furthermore, Senator Brault, the floor sponsor of Article XI at the 1969 special session, referred repeatedly to section 1 as a "mandate."

This statement of public policy becomes a mandate directing all arms of the State to consider the impact of proposed actions upon the Commonwealth's environment. Section 1 does not tell an agency how to choose between conflicting factors such as economic development versus environmental impact. But it does require that, along with whatever factors (economic or otherwise) which by statute bear on an agency's decision, the agency must examine and weigh environmental consequences. Moreover, the agency must do so conscious of the fact that it is no ordinary statute, but the Commonwealth's fundamental law, which declares public policy as to natural resources and environmental amenities.[2]

This language, approved by the legislature in 1969 for vote by the people, predates the 1970 National Environmental Policy Act, which requires the same decision analysis before acting. If much of what Professor Howard is explaining here strikes some of the Commonwealth's agencies as unfamiliar—agencies whose mandate is not environmental but rather economic development, transportation, or energy, for example—I would simply point out that this analysis has been available for almost fifty years now and is even less well known than Article XI itself.

As for the environmental agencies, Professor Howard gives them a firm foundation as well:

Such a constitutional declaration gives additional backing to the actions of state agencies whose statutory mandate it is to police the environment, such as the State Water Control Board and the Air Pollution Control Board. The special place the constitution gives to environmental quality among the values which guide public action will be relevant in defending an agency's action against the charge, say by one who is ordered to conform to antipollution standards promulgated by an agency, that the agency has acted too vigorously or imposed standards which are too stringent.

The overwhelming majority of people want these agencies to do their job, to protect Virginia's environment. If the environmental agencies need to justify their legitimate actions to protect the Commonwealth's atmosphere, lands, and waters from pollution, impairment, or destruction, I recommend they read this constitutional analysis again. As with the "elimination of discharges" language in the Clean Water Act, it is time to carry out Article XI's mandate routinely, not occasionally.

The Supreme Court of Virginia has taken a different view from that of Professor Howard, the author of Article XI. In the case of *Robb v. Shockoe Slip Foundation,* issued January 18, 1985, the Court stated that Article XI was not self-executing and further that "the only purpose for adding (section 2) to Article XI was to instruct the General Assembly to enact statutes whereby the public policy declared in section 1 could be executed."[3] The framers of Article XI had a different view than the court, but the court has the final word for now.[4]

In the state of Montana, the Supreme Court took a more comprehensive view of the right to a clean environment.[5] In a 1999 decision, *Montana Environmental Information Center v. Department of Environmental Quality,* the Montana Supreme Court ruled unanimously that Montanans' constitutional right to a clean and healthful environment (Article IX, Section 1) is a fundamental right and one that is intended to be preventative. In an opinion by Justice Trieweiler, the Montana Supreme Court held that "our constitution does not require that dead fish float on the surface of our state's rivers and streams before its farsighted environmental protections can be invoked" and concluded that "the delegates' intention was to provide language and protections which are both anticipatory and preventative," establishing that the right is preventative in nature.[6]

Virginia's Article XI uses different language than Montana does, but in stating that the Commonwealth's policy is to "protect its atmosphere, lands and waters from pollution, impairment or destruction for the benefit of the people of the Commonwealth," it is reasonable to conclude that such protection goals mimic Montana's preventative language. At a minimum, people in Virginia can expect conservation of natural resources and protection of air, land, and water. Article XI of Virginia's constitution became a touchstone for environmental conservation in Virginia going forward because of this policy. Many efforts to improve Virginia's environment initiated during the Holton administration, and later by the Virginia Environmental Endowment and the many nonprofit organizations it helped develop, can be traced to this historic mandate.[7]

As a fundamental statement of Virginia's public policy on the environment, Article XI remains less well known than it ought to be—particularly to those who swear to uphold it, many of whom have little idea it is in the state constitution. Although Article XI applies to state agencies, it also applies to local government decisions concerning, for example, land use and water quality matters, because local officials also take an oath to uphold the constitution of Virginia and Article XI.

# 6

# Permission to Pollute

## THE CLEAN WATER ACT

DURING THE spring of 1972, Senator Edmund Muskie of Maine held hearings on the proposed Federal Water Pollution Control Act. He was also running for the Democratic presidential nomination that year, so clean water got a lot of attention. In the imposing Senate committee room, the Senators were arrayed around an elevated semicircular platform to hear testimony and mark up the draft bill. These politicians had heard the Earth Day messages for three years in a row: people wanted clean water, and they expected the Congress to do something about it.

It was my first visit to the impressive halls of the storied Russell Senate Office Building. The Committee on Public Works was in session, and the room was filled to overflowing with clean water advocates, Senate staffers, lobbyists, television cameras, and reporters. The energy and enthusiasm were impossible to ignore. It felt like a historic moment.

I accompanied our governor, Linwood Holton, and helped write his testimony to the Committee, the gist of which was, "Clean water is a great idea; do your best, just don't supersede our excellent Virginia water law." The Committee heard many speakers that day. A few objected to parts of the bill, but most of the speakers and Senators spoke to its merits and called for its passage. Clean water was the goal; eliminating pollution discharges was the strategy. That day, in the US Senate, there was no mistaking the intent of the legislators to clean up the nation's waters.

The United States Congress voted overwhelmingly to enact the Federal Water Pollution Control Act in October of 1972, just weeks before the presidential election. President Nixon, who supported the environment in principle, vetoed the act on October 17, 1972, saying it contained four times more spending than in his budget bill. The veto was overridden by the Senate that same day and by the House the next day. The FWPCA72 became the law of the land.

Nothing illustrates the gap between the promise of an environmental law and the realities of implementation better than the Clean Water Act's

National Pollution Discharge Elimination System (NPDES). The goal of the Federal Water Pollution Control Act of 1972, which was amended in 1977 and renamed the Clean Water Act, is "that the discharge of pollutants into the navigable waters be eliminated by 1985."[1]

When the act's regulations were initially drawn up, there was disagreement about how pollution should be eliminated. Government regulators were creating a new regulatory process, and the regulated industry people were saying, in effect, "for Pete's sake, just tell us what you want, and let us figure out how to do it." They lost that fight. The "command-and-control" argument won the day, because, to oversimplify, the government didn't trust industry to do the right thing.

Although the Federal Clean Water Act prohibits any person from discharging pollutants into surface waters, in practice all persons are instead required to obtain a National Pollution Discharge Elimination System permit prior to the discharge of pollutants into surface waters through a point source.[2] To be clear, that means they get a permit to discharge what the law says to eliminate. While the worst of the discharges from fifty years ago have been reduced, the permit system remains in place. Thus, in Virginia and many other states, implementation of the law turned it into a program for issuing permission slips. Instead of an elimination system, as the title of the program states, it is a discharge *permission* system. Pollution permits remind me of a parent telling their kids, "Don't color on the wall, but if you do, just use these three crayons."

Permits do contain limits on how much poison can be discharged. They also specify reporting requirements and contain other provisions to ensure that the discharge does not degrade water quality or harm public health. The permit translates general requirements of the Clean Water Act into specific provisions tailored to the operations of each entity discharging wastes into the waters of the United States. States also establish water quality standards that are intended to limit damage to water quality from permitted discharges. Former director of Virginia's Department of Environmental Quality David Paylor, who spent more than forty-five years in distinguished state service protecting the environment, stated at the Environment Virginia conference in 2021 that the permit system is designed to be sure there is no harm to people and the environment "and/or to keep it to the point where it is having no negative impact."

Nonetheless, the permit system continues a practice—discharges—that Congress intended to eliminate.

# 7

# The Clean Water Act in Virginia

PRESIDENT NIXON made a difference in the nation's environmental quality. He established the Environmental Protection Agency (EPA) in 1970, continuing the Republican party's long history of support for environmental conservation, which dated back to President Teddy Roosevelt. He signed the National Environmental Policy Act on January 1, 1970, and set up the Council on Environmental Quality. He signed the Clean Air Act of 1970, the Marine Mammal Protection Act of 1972, and the Endangered Species Act of 1973. And his EPA banned the use of the pesticide DDT in 1972, an action that has helped to restore the bald eagle population in Virginia and around the country. The environment was a national priority in the 1970s.

Linwood Holton served as the sixty-first governor of Virginia. In his inaugural address, Holton said, "This administration [is] . . . determined to make the quality of our air and water . . . top-priority public concerns." In the early 1970s, thanks to bipartisan efforts in the Congress and leadership from President Richard Nixon, an unprecedented amount of federal funds became available to states and localities for sewage treatment plant construction. The federal share of this new grant program was 75 percent; the state had to chip in 20 percent, and localities only 5 percent. It was a major boost for clean water, and Holton made sure that Virginia got its share.

The responsibility for regulating water pollution in Virginia belonged to the State Water Control Board (SWCB). Governor Holton replaced all the members of the board, and he appointed a nuclear engineer from Northern Virginia named Noman M. Cole Jr. to serve as chair. The new SWCB was extraordinary in its commitment to clean water, and it used Virginia's water control law aggressively to limit discharges into state waters and to enforce violations of the law.

Cole had been personally selected by Admiral Hyman G. Rickover, the father of the nuclear navy, to be a part of his team during the 1960s. The admiral trained Cole well in the arts of command, control, and communication. Rickover was a notorious taskmaster; in selecting people to serve in his command, he emphasized excellence, determination, intelligence, and persistence. Noman Cole fit right in. Cole, who was thirty-seven when he became the SWCB's chairperson, was interested only in results, not excuses.

Cole served as chair from 1970 until 1974. He lived on Mason Neck in Fairfax County. A very intelligent man, he had a plainspoken way of expressing himself rarely heard from public officials. Cole had a high-pitched voice when he got excited or exasperated, something that happened often when he presided at State Water Control Board meetings and had to listen to long-winded presentations from local water district engineers. Cole could be relentless in his questioning, leaning aggressively forward into his microphone for greatest effect, and he barely contained his disdain for any lack of preparation or for anyone unable to answer questions to his satisfaction. He would routinely rip into unprepared witnesses who came before the board.

Cole brought his can-do confidence to his part-time volunteer job as chair of the seven-member SWCB, viewing it as a target-rich environment for controlling if not eliminating water pollution. He and his colleagues on the board believed and acted strongly in defense of clean water.

For example, Cole was outraged by the pollution of the Potomac River from sewage discharges in the Washington, DC, metro area, both in Virginia and in the District of Columbia. The Potomac was his waterfront. Cole read the state water control law and said, "We are going to enforce this law!"

New federal and state money had been designated to fix the municipal discharge of sewage that localities would collect, flushed from toilets into big pipes that carried the sewage to a plant designed to treat it and reduce the level of harm that discharges caused to receiving streams' water quality. Yet, in the early 1970s, Fairfax County had a sewage treatment problem. Residential growth was rapidly turning farms into subdivisions. Each home in each subdivision required a sewer hookup, which in turn was connected to a larger pipe that eventually was connected to a sewage treatment plant. County engineers and local developers were used to the State Water Control Board accommodating requests for additional sewer hookups so that the subdivisions could be approved by the Board of Supervisors.

Cole and the rest of the board recognized a serious problem: the large increases in the amount of sewage being deposited into the treatment plants were overwhelming the plants. The plants no longer had enough capacity to handle the flows, resulting in increased water pollution. The SWCB wanted to increase the plants' capacity by adding higher levels of treatment. It also wanted to build new plants, designed to provide a level of treatment referred to as "tertiary treatment," because the existing "secondary treatment" plants could not adequately treat the loads they were receiving before discharging municipal pollutants into state waters. Cole wanted tertiary treatment to be the new standard, and the federal government at that time had plenty of money for constructing such plants.

He said, "Look, you got a lot of big pipes coming into a small plant. You don't have the capacity to manage all that waste. It is not being treated completely, and it pollutes the water. Now, what are you going to do about that?" They replied that, because of all the new residential development in Fairfax County, they had little choice but to allow more sewer pipes to be constructed.

Cole and the board in effect said, "No, you can't allow more sewage connections until you also increase plant capacity." The SWCB had had enough of land development's effect on water quality.

The state water law provided that if the amount discharged from new housing and commercial developments exceeded the treatment plants' capacity, then the SWCB could direct the locality to stop issuing development permits. The law thus allowed the board to impose sewer hookup moratoriums on local governments that permitted more hookups than their sewage treatment plants could handle. It was a little-known power to regulate the connection between land use and water quality.

The board was willing to help localities construct new or upgraded sewage treatment capacity. The Environmental Protection Agency (EPA) was flush with federal funds to distribute to states and localities that wanted the money. Local governments had to put up only 5 percent. Cole also developed solutions. He was the principal author of the 1971 Occoquan Watershed Policy, which prompted creation of a sewage authority there, as well as of a world-class treatment plant.

The difference between this Holton board and other water boards I have observed is that the Holton board was aware of their duty to uphold their oath to the new Virginia constitution, and they vigorously enforced the state water law. Holton expected nothing less, and they knew that he had their back.

Despite earlier warnings, in 1973 Fairfax, Hampton Roads, and Roanoke all exceeded their plants' capacities. The SWCB then imposed a sewer hookup moratorium in all three places. The reception the board received for this courageous act to enforce the law was colder than a trout stream in March.

Localities were stunned. They complained to their elected representatives in the House of Delegates and the State Senate in Richmond, and to the new governor, Mills E. Godwin Jr., when he took office in 1974 for his second term. In 1974 the General Assembly repealed the section of Virginia's water law that gave the board the authority to impose such moratoria. The leadership pioneered by Noman Cole and his board was squelched.

Running to the General Assembly to restrict the decision-making capacity of environmental citizen boards continues on a bipartisan basis today. As recently as February 2022, the Senate of Virginia voted thirty-two to eight to strip the air, water, and waste management boards of their power to issue or deny permits, instead giving that authority to the Department of Environmental Quality staff. Citizen boards that once controlled and supervised the regulatory agency and hired the top staff have now been stripped of those authorities. Peggy Sanner, Virginia director of the Chesapeake Bay Foundation, in commenting on this legislation, said, "Citizen boards ensure that Virginians have a meaningful voice in shaping the rules that are intended to protect our air, water, lands, and communities."[1] Ms. Sanner added, "The experts weigh in with their advice, but the environment is something that belongs to Virginians, and in fact the Virginia constitution makes it clear that the environment is for the benefit of all people. . . . The air board and the water board make sure that decisions affecting our environment are not made behind closed doors—agency representatives to industry representatives, but in fact involve the people."[2]

Another, more constructive solution to these kinds of problems would be to focus less on limiting development or on repealing inconvenient laws and instead, as we will see in a later chapter, to allocate more money to build or expand sewage treatment plant capacity and solve the real problem, which is water pollution.

The idea of preventing pollution and eliminating waste discharges in the first place is still the official policy of the Clean Water Act, but permission is what occurs. Nevertheless, many corporations have figured out that it is in their own self-interest and better for their bottom line—and better for the environment—to get maximum efficient use of their material resources in creating their products rather than to continue wasteful, inefficient processes that can lead to discharges, pollution, and in the worst cases, pollution disasters, prosecutions, and expensive costs.

In the late 1990s, during Governor James Gilmore's administration and under the leadership of Lieutenant Governor John Hager, the Virginia Governor's Environmental Excellence Awards were established in partnership with the Virginia Manufacturers Association. I served on the panel with Governor Hager to select the award winners. It was a wonderful and positive experience.

There were four categories of rewards for small businesses, such as Trex in Winchester, and for large businesses, like DuPont in Waynesboro. Each corporation could choose to submit documentation of either ongoing major programs, smaller one-time projects, or both kinds. Year after year, companies would come up with environmental innovations. For example, some companies switched from a toxic material to a nontoxic alternative, eliminating pollution from their process entirely. Other companies aimed at improving their efficiency to the point where they could recapture every emission and discharge nothing.

There were two criteria for the excellence award. The first was that the innovation had to eliminate pollution—eliminate!—to the point where it was nontoxic or not discharged at all. The other criterion was that they had to save money by doing so. Both were important. And that's how we wound up with dozens and dozens of Virginia companies submitting applications each year showing that they had done both of those things. There were first- and second-place awards, and we also honored other companies' accomplishments with an honorable mention award for eliminating discharges while saving the company money by doing so. Over the years, dozens of Virginia companies have met this environmental excellence standard and significantly prevented pollution. It is a straightforward calculus: becoming more efficient or changing the substances from toxic to nontoxic almost always reduces both costs and pollution. Elimination is possible.

Today, the awards continue in a different way as part of the annual Environment Virginia conference. The Department of Environmental Quality also runs its Virginia Environmental Excellence Program (VEEP), which "encourages and assists facilities and organizations that have strong environmental records to go above and beyond their legal requirements. The program has grown steadily since its inception and currently has over 450 members."[3] I hope that the original criteria for eliminating polluting discharges while improving the bottom line have endured.

For much too long, pollution meant profit, at least in the short term. The dire consequences to Allied Chemical Corporation from the Kepone case should have sent a message about that fallacy decades ago. The companies and engineers that were recognized in the Governor's Environmental Excellence Awards program are proof that prevention is both possible and profitable, not to mention less risky. Prevention of pollution is also good public policy, as mandated by Article XI of the Commonwealth's constitution. There is no constitutional right to pollute in Virginia.

# 8

# The Clean Water Act

## A CASE STUDY AND NEW CHALLENGES

THE ENDOWMENT decided to investigate how Virginia was managing its National Pollution Discharge Elimination System program and to learn what those permission slips were allowing to be discharged into the James River.

I recall a day in 1979 when the seven-member Virginia Environmental Endowment board was seated around a conference table at a quarterly board meeting in Richmond. We were discussing the condition of Virginia's rivers and how little information there was available to the public about water pollution and the sources of it. Sydney Lewis, who did not speak up often, looked at me from across the conference table and asked me, "Mr. McCarthy, do we have any idea what is being discharged into the James River?" Admiral Bullard noted that there was plenty of pollution in the Hampton Roads area. And Judge MacKenzie bluntly said, "The Elizabeth River is a mess."

As I mentioned in a previous chapter, in 1976 Virginia enacted a Toxic Substances Information Act, which was followed later that year at the federal level by a comprehensive Toxic Substances Control Act (TOSCA). The Virginia law required companies manufacturing and using chemicals to disclose the details of their emissions and discharges to the state government. The law also required that the information be kept confidential and not disclosed to the public. Therefore, we did not have access to information on what was being discharged into the James. With the Kepone discovery only a few years past, it seemed to the board that someone ought to find out. As we will see, asking the question was the easy part.

By the time it was discovered in 1975, Kepone had been discharged into the James River for many years. However, nobody had known about it, because at the time, the technology for detecting toxic substances in water could identify only what a researcher knew they were looking for. If they didn't know what they were looking for, or if they just weren't looking, they couldn't find it.

For this reason, the idea that the state was giving permission to discharge all manner of substances into rivers and streams struck the board as odd. Permit applications to discharge waste into state waters are supposed to describe in detail exactly what will be discharged, if permitted. The SWCB staff reviews the applications and negotiates the details of what the permit will allow to be discharged. In that case, both the applicant and the regulators presumably know what will be discharged. The public does not—and, as we will see in a later chapter, did not at that time—have any easy way to find out.

One board member, George Yowell, leaned forward over the conference table. "You mean, no one really knows what's being discharged routinely into the James?" he asked, wide-eyed.

"No," I said. "But we might be able to examine the discharge permits issued by the state of Virginia."

Tom Wolfe, referring to the Clean Water Act's mandate to eliminate discharges, wryly observed, "Nobody's eliminating anything."

Bill Cummings, our board chair, who was still serving as the United States Attorney for the Eastern District of Virginia, settled the matter: "I think we need to look at those permit files at the State Water Control Board, and see what's really going on, what's being discharged."

The following week, I made a few calls to national environmental groups to see which of them might have the legal and scientific staff to be able to carry out a review of the state water discharge permits for us. At that time in Virginia we did not have the kind of expertise required for such a task. The National Wildlife Federation (NWF), based in Washington, DC, had just the right people: lawyers and scientists who were familiar with the requirements of the federal law's National Pollution Discharge Elimination System (NPDES) program. I asked them to send us a proposal to do the work.

The proposal from the NWF set out to identify and rank potentially hazardous chemicals being discharged into the James River and to evaluate the implementation of the NPDES system and the Commonwealth's Toxic Substances Inventory law. The goal of this work was to help regulators and citizens understand what was going on and set priorities for the management of toxic substances to prevent future Kepone-like problems.

We awarded the grant, hoped for the best, and looked forward to seeing what they would find. This was probably the first time an outside group had taken a detailed look at what kind of poisons were being discharged into the James River—perhaps any river.

What started out as a routine investigation uncovered disorder in the permit operation at the State Water Control Board. After only a few months working on the study, Ken Kamlet, the principal investigator, called me:

"You're not going to believe what we're finding; it's a mess. Most of the permits, which are issued for a five-year period, are out of date and haven't been renewed or updated in years." The system was a shambles, resulting in a lengthy backlog of requests for permit renewals and expansions. Some of the information was almost ten years old. We had assumed, after Kepone exposed some weaknesses in the Commonwealth's tracking of discharges, that by now the permits would be well documented and up-to-date, with details of discharge materials and amounts, but they were not at all.

NWF issued a draft of their report, which drew howls of protest. Several of the companies who held the permits reviewed the first draft and criticized NWF for using outdated information in compiling its findings. When they realized the problem was that the Water Control Board had not incorporated the companies' latest information to issue new permits, they were flabbergasted. The companies had submitted the information, but the agency was years behind in processing the information and writing new permits. The report's findings eventually led to a major shake-up in the permit operations at the agency. The agency is now known as the Department of Environmental Quality, after a reorganization in 1993 that merged the staffs of the air, water, and waste boards.

Yet the permit system survives to this day. On its website, DEQ describes the permit system this way: "The Clean Water Act established the National Pollutant Discharge Elimination System (NPDES) program to limit pollutants getting into streams, rivers and bays. DEQ administers the program as the Virginia Pollutant Discharge Elimination System (VPDES)." The agency issues permits for all point source discharges to surface waters, discharges of stormwater from Municipal Separate Storm Sewer Systems (MS4s), and industrial discharges of stormwater. The idea of eliminating discharges has gotten completely lost.

Overcoming the inertia of the state permit writing process was always going to be a major challenge. It is true that some of the worst of the discharges have been reduced, but the idea of elimination or prevention remains a goal without either a plan or annual objectives. The current permit program issues thousands of permits every year.

Think about that.

## Total Maximum Daily Load

Another concern with the CWA's implementation is the Total Maximum Daily Load (TMDL) requirement. The idea behind the TMDL approach is to reduce, not eliminate, the poison runoff that is discharged from farms, fields, forests, and urban and suburban sources, to levels that might protect the Commonwealth's waters from pollution, impairment, or destruction. The TMDL requirement states, in effect, that what is getting into the rivers and streams must be limited to what the water can safely absorb. The approach depends on dilution, not elimination, of pollution runoff.

In the mid-1980s, we made a grant to the Natural Resources Defense Council (NRDC) to look at "nonpoint sources." There was a certain amount of focus on Virginia in the resulting book, because the writing of it was made possible by the VEE grant. NRDC was surprised by what they found, which was that no one at the SWCB was doing anything to implement this part of the Clean Water Act.

Paul Thompson, Robert Adler, and Jessica Landman led that effort. That grant resulted in the publication of *Poison Runoff*,[1] which at the time was the most comprehensive look at the Clean Water Act and the states' efforts to implement it. It's still an excellent primer on the act and the "nonpoint source" problem that the act's TMDL section is intended to solve. NRDC told me: "Look what we found in the CWA!" It was the previously little-known Section 303, which calls for the TMDL solution: "We found this section that nobody is implementing."

We called this discovery to the State Water Control Board staff's attention. They were not aware of it and had no experience with it. But it was in the Clean Water Act and therefore their responsibility. The section called for the development of TMDL plans; no one at SWCB had ever done a TMDL plan.

To help remedy this situation, VEE made a $66,194 grant in 1990 to the University of Virginia School of Engineering to do a pilot TMDL study on Muddy Creek in the Shenandoah Valley. The aim of the research was to answer the question "How does one do a TMDL study? What does a TMDL plan look like?" The UVA engineers and community experts sought to demonstrate what to do and how to do it.

Not only did UVA bring in engineers and scientists, but they also included a community component of engaging in public outreach and enlisting support for this new approach to water quality management. It took a good year or more, because UVA did the community involvement part as well. Both aspects of the plan were necessary and equally important. When finished, they turned in their report and "how to" instructions over to the SWCB staff. The initial reception was lukewarm at best, but after further encouragement, SWCB began to develop TMDL plans.

Over time the TMDL approach has become well established and is now being applied to the entire Chesapeake Bay, the largest and most complex estuary in the country. In 2009 the Chesapeake Bay Foundation (CBF) sued the EPA over its failure to uphold the Clean Water Act. The 2010 settlement of that lawsuit resulted in a TDML plan designed to restore water quality in the Bay by 2025 by allocating pollution reductions to each of the six Bay states and the District of Columbia. Each jurisdiction has been required to create and implement plans including milestones to achieve those reductions, resulting in a "blueprint" that can be used on other TMDL plans. In September of 2020, CBF and its partners took the next step toward requiring the US Environmental Protection Agency to uphold the Chesapeake Clean Water Blueprint and ensure that pollution-reduction goals for the Chesapeake Bay and its rivers and streams are met. Many people working over a period of decades deserve credit for getting the government to pay attention to the TMDL section.

# 9

# Environmental Policy in Virginia

## HOW THE HOLTON ADMINISTRATION LAID THE FOUNDATION FOR ENVIRONMENTAL IMPROVEMENT

LINWOOD HOLTON set Virginia's modern environmental era in motion. A Harvard-educated lawyer, Holton, at only forty-six years of age, toppled the conservative Byrd Democrats in the 1969 election to become the first Republican governor since Reconstruction.

At his inaugural address, Holton laid out a progressive set of priorities that promoted civil rights, repudiated Virginia's massive resistance to integration in public schools, and elevated the state's responsibility to protect the environment. Holton famously declared that "the era of defiance is behind us." He further proclaimed, "We know our lakes are dying, our rivers growing filthier daily, our atmosphere increasingly polluted . . . This administration [is] . . . determined to make the quality of our air and water . . . top-priority public concerns."

Governor Holton assumed office just as the modern environmental movement was being born. President Nixon had signed the Clean Air Act, created the Council on Environmental Quality, and established the Environmental Protection Agency. It was 1970, and plans for the first Earth Day in April were well underway. Virginia was about to adopt a new constitution stating that it was the Commonwealth's policy to "protect its atmosphere, lands and waters from pollution, impairment or destruction." Governor Holton signed an Executive Order creating the Governor's Council on the Environment and notified his executive assistant, Rodger Provo, to find someone to run it.

I had recently moved to Virginia after being honorably discharged as a young captain from the United States Air Force, having served in the Nuclear Weapons Division of the Air Force Weapons Laboratory at Kirtland Air Force Base in Albuquerque, New Mexico. I wrote to the new governor telling him I was a new resident, congratulating him on his election and offering to be of help. I enclosed my resume, not expecting he would see it. My letter landed on Rodger Provo's desk. A week later, he called from the governor's office. My job objective had matched their position descrip-tion for the executive director of the new Governor's Council on the Envi-ronment. "Would you be interested in interviewing for the position?" he asked. I had never been to Richmond, never lived in Virginia, and never met a governor, but the job sounded interesting.

During the summer of 1970, Rodger scheduled a series of meetings for me with three of Holton's closest advisors. Each one greeted me warmly and asked me about my background, interest in the environment, and other relevant matters. They were keenly interested. I explained what I'd learned about current environmental issues from my research over the past year, detailing the people I'd met and the big-picture conclusions I'd drawn. I probably sounded a bit naïve, but I saw a great opportunity to solve environmental problems, and my enthusiasm was genuine.

After I'd completed the interviews, Rodger invited me to meet the governor himself. As I merged onto I-95 South from McLean, bound for Richmond on that sunny summer morning, it struck me that I had come a long way from nuclear engineering in Albuquerque. Ninety minutes later, I was searching for a parking place near the state capitol.

As if I wasn't already intimidated, Thomas Jefferson's neoclassical state capitol building in Richmond inspired both awe and nerves. Rodger Provo met me at the entrance. He was even younger than I was, and he greeted me with a broad smile before guiding me across the capitol grounds and up to the governor's private office on the third floor. We ascended in an elevator so tiny that the two of us about filled it. It seemed as though it could be old as the historic capitol building—designed by Thomas Jefferson and com-pleted in 1788—except that public passenger elevators were not available until Elisha Graves Otis invented the first one with a safety brake in 1854.

After some of the generals' offices I had visited while in the Air Force, I was surprised at how small the governor's office was. It was almost in-timate, although beautifully furnished and decorated, with a lovely view to the north side of the capitol. A handsome and lively man, wearing a dark blue suit, a white shirt, and conservative tie, Holton thrust his hand out and pulled me into a vigorous handshake. The sign on his desk read, "Today is opportunity day. Do something." I liked him right away.

Outside the governor's third-floor window, on the north lawn, there was a preacher using a megaphone to rally a small crowd, working his way into a rhetorical lather and expressing his outrage at this "liberal gover-nor" who dared to promote racial integration in Virginia's public schools. Holton pointed him out and laughed, "He's not a fan."

Just a week before, Holton had escorted his daughter Tayloe into the downtown Richmond public school closest to the executive mansion. Holton could have sent his daughter to the finest private school in town. Instead, he backed up his principles with action. He believed in integration, and he set the example with his own family. Not all Virginians approved, but it sent a powerful message. The photo of Holton and his daughter entering the nearby public school appeared in newspapers and magazines all over the country.

Rodger introduced me to John Ritchie, Holton's chief of staff, and the four of us sat down to talk. I was used to business meetings without preambles; in the Air Force we quickly got down to business. Chatting with these three men, I quickly concluded by the friendly banter that in Virginia one "visits a while" before getting down to business. I knew a lot about the environment, but not much about Virginia.

Moving on from the small talk, we spent a lively few minutes discussing my interest in helping the governor realize his environmental goals and learning how much fun he was already having as governor (a lot). Then, just like that, Governor Holton asked, "When can you start?"

I was so pleased and excited that the rest of the conversation was a blur. I was hired! Delighted and more than willing, I said, "Next Monday!" We shook on it all around, and leaving the office I turned to Rodger and said, "I just got hired?" He laughed and said I would love it. I was twenty-seven years old and my new state was so unfamiliar to me that my most prized possession was a Virginia state highway map.

LINWOOD HOLTON loved being governor and was an enthusiastic salesman for all things Virginia—its educated workforce, its natural beauty, its openness to business investments, and perhaps the most fun, Virginia's new tourism slogan, "Virginia is for Lovers." When he was out representing the Commonwealth at meetings, he always had a plentiful supply of "Lovers" campaign-style buttons and bumper stickers to hand out at every opportunity.

I was hired to serve as the executive director of the Governor's Council on the Environment, which he had created by Executive Order soon after taking office. My job involved developing new policies, coordinating a multitude of state environmental conservation agencies, and developing the governor's environmental agenda.

I initially shared a small office across Ninth Street from the capitol with Rodger Provo, the executive assistant who was responsible for my hiring.

That was a good thing, because Rodger had his finger on the pulse of what mattered to the governor. As a former reporter, he was used to gathering and using information and was a great help to me starting out that September of 1970, telling me about the agency heads I would be working with and how best to get along with them: "Go to their offices, meet them, learn what they're doing, where the governor can help, not just what they can do for you."

At first, I felt more like a junior Air Force officer trying to enlist the support of too many generals who were used to doing things their own way. But that feeling passed. Most of these men—there were no women environmental agency heads back then—were helpful once they overcame their initial suspicion of what the governor was up to, and they concluded that we had a mutual interest in improving Virginia's environment.

When I joined Governor Holton's administration, I had no idea how significant his term would be in the history of Virginia. In the first months of the Holton administration, I thought it was about solving problems, the way an engineer thinks: identify the problem, analyze it, develop a solution or two, and get on with it. I remember Governor Holton telling me that progress was incremental, but in Virginia it was time to make the environment a priority. Now.

It took me a while to understand legislators who wouldn't agree with our proposals. The environment was, after all, a popular cause. I never said to them, "How could you be against environmental protection?" They could find all kinds of procedural ways to be against any proposal if they chose to.

I had become accustomed to making presentations when I was in the Air Force. There, the arguments usually won the day, because they were premised on science and fact, and in the nuclear weapons world I lived in, the world was going to go to hell if we didn't get it right. Political horse trading is not so simple. There are many factors involved in why anybody would support you or oppose you. I really was naive, because I'd never dealt with governors and legislators before. Political considerations must be understood and accommodated. Some legislators might be with you on one bill but might not be with you on another—nothing personal, just reflecting a constituency, a party line, or something else. It took me a while to grasp this reality. Elegant solutions—the stuff of engineering—were not that common. A good victory was usually a good compromise.

Governor Holton was an effective governor because he focused on results. As an example of how this worked: one day I had asked to see him to describe a problem, hoping that discussing it with him might give me direction about what to do about it. He listened, and after a minute, growing impatient, interrupted me to say, "Jerry, your job is to bring me solutions, not problems."

Once again, I noticed the sign on his desk. He was always busy doing something. He set a great example. He went on—as if "bring me solutions" wasn't enough to absorb!—"You know what my job is? I point! When something needs doing, I point—at you, or John (Ritchie) or Rodger (Provo) or Ed (Temple) and expect you to get it done."

Of course, figuring out what was to be done was one of his great strengths as a leader. Then he'd point to someone to do it, and it got done. He was decisive. When you did bring him a solution, having duly been pointed at, he'd decide and move on to the next matter.

Linwood Holton, in all my dealings with him, displayed a clear sense of the possible in politics, particularly when it came to negotiating with the overwhelmingly conservative Democratic majority in the Virginia legislature. One of his major accomplishments was to create a cabinet of top-level administrators to help the governor manage the sprawling bureaucracy contained in the more than 125 agencies of the state government, all of which were directly overseen by the governor. The idea of a six-member cabinet was a product of a legislative study commission, but Holton was an enthusiastic backer of this much-needed change. Of course, the idea was immediately attacked as yet another level of government bureaucracy that was going to cost the taxpayers more money and not improve matters. His calculation in the 1972 General Assembly session that he could get a cabinet of six members approved by the legislature (not a seventh, such as was hoped for in the area of natural resources) proved correct: "Jerry, if I add a seventh cabinet member, the whole thing will go down in flames." By the slimmest of margins he succeeded in establishing the cabinet in Virginia's executive branch. It has expanded dramatically in size since 1972 and now includes a Secretary of Natural Resources thanks to the leadership of Governor Gerald L. Baliles.

The Governors' Council on the Environment was initially chaired by the governor. He appointed the same three advisors I had met with previously. Former state senator FitzGerald Bemiss of Richmond had been a longtime legislative champion of land conservation in Virginia and helped convince Democratic conservatives to support Holton as a candidate for governor. John Hanes lived in Northern Virginia and was an investment banker on Wall Street; he also chaired the finance committee for Holton's campaign. Charles Williams was a member of the World Future Society and a member of President Nixon's White House policy staff. These three citizen appointees comprised the executive committee, set the agenda for our work, and were the people I worked the most closely with.

The governor filled out the Council membership with the heads of the state environmental agencies, the state health commissioner, the highway commissioner, and the attorney general.

As the partial responsibility of several agencies, issues like the use of land and its effect on water quality received little attention at the state level, with nobody taking the full measure of their impact on the environment. My job was, as John Hanes, the Council's vice chair, put it, "to get all the agency horses running in the same direction at roughly the same speed." In my first few weeks, I scheduled one-on-one meetings with all the agency heads who served on the council. Most agency heads were polite and happy to meet me, but while they were all for coordination, none of them wanted to be coordinated! One longtime agency head told me he had "seen a lot of young people come into government . . . and soon they all were gone." I politely assured him that I took the job of protecting the environment seriously and that the governor would value his cooperation. Over the next few decades, in various roles, full-time or as a volunteer state board member, I served nine governors.

Starting a new agency from scratch was exciting and full of anxiety as well. I had "stage fright" at being suddenly in the highest reaches of state government in Virginia—a state new to me—working for the governor, advising him, coordinating the implementation of environmental programs, and developing his environmental agenda. Rodger Provo and the governor's close friend and vice chair of the Council, investment banker John W. Hanes Jr., were helpful in serving as my guides to all things Virginia and building my confidence.

All the environmental agency heads were nominally supportive, but none of them wanted to budge from business as usual, and Linwood Holton was all about abandoning the status quo. After ten or twelve weeks of meeting with agency heads and holding a couple of Council meetings, it seemed as though we weren't accomplishing much. I was frustrated by the seeming lack of progress and not quite sure what to do.

Even though Rodger Provo and I shared an office, we weren't usually there at the same time. One morning, sipping black coffee and in an equally dark mood, I expressed my frustration to Rodger: "You know, sure, they're all in favor of what we're trying to do, but nobody wants to make anything happen. Sometimes I feel like these older agency heads are just patting me on the head and saying, 'There, there, it's going to be fine.'"

As I had learned to expect by now, Rodger was both wise and blunt in his analysis: "Jerry, nobody in government in a leadership position is going to last long if they don't have a constituency to rally around and support a shared cause. You risk sinking into irrelevance if you don't think of something quick." The Council had the Governor's strong support, which is vital and necessary, but it was not sufficient for the tasks we had been assigned.

In a letter addressed to John Hanes, the Conservation Council of Virginia expressed its collective upset with the lack of progress being made by Governor Holton to fulfill the promise in his inaugural address to protect Virginia's environment. Despite the belligerence of the letter's tone, Rodger and I agreed they were making some good points. Promises made ought to be promises kept, although less than a full year in office was hardly enough time for the kind of results they expected.

I think it was Rodger who first came up with the idea of holding public hearings about the environment. "We need to get out there," he said. The previous April, people all over Virginia and the nation had celebrated the first Earth Day, and the environment was a hugely popular subject. We needed to capitalize on that untapped support to fire up our agenda.

In January of 1971, we advertised a series of public meetings to hear what the public thought about environmental matters and what we were supposed to do about it. The meetings were held in Charlottesville, Norfolk, Northern Virginia, Richmond, and Roanoke on January 25, 26, 28, and 29 and February 1, 1971. Large turnouts greeted us wherever we went, and in Northern Virginia we had to return for a second night to accommodate all the people who wanted to be heard.

The members of the Governor's Council on the Environment had never come together as a group of top Virginia officials in a public hearing to listen to what the people had to tell them, sometimes bluntly, to their faces. These hearings were the first public hearings on environmental improvement in Virginia, a routine practice now but a new idea in 1971. The reluctance of some Council members to attend was in part because nothing like this had ever been done, but each was present at every hearing. It was not always easy for them to listen to criticism of their agencies, but they did it. Perhaps in doing so, they saw that this governor was on to something with his emphasis on environmental quality and, even better, that they had better get on the right side. Council members became more cooperative when they discovered the popularity of environmental quality with the governor and the public, and how it might benefit their agencies.

Hundreds of people from throughout Virginia gave us an earful. Some, no doubt, were interested citizens attracted to the idea of telling top government officials what they thought. Others spoke on behalf of a variety of local groups and neighborhood associations or were people holding local office or running in the next election, and some clearly identified as conservation group representatives. In each of the halls where the public hearings were held, the seats were filled and the energy in the rooms was palpable. The well-behaved crowds applauded each speaker enthusiastically. Folks were ready to speak their minds. A lot of pent-up frustration surfaced when they were given this opportunity. Most speakers were polite, positive, and persuasive in expressing strongly held views:

"Can the world be saved from the ecology crisis? Of the several million species of organisms which inhabit our planet, only one can pose such a question, and only one will be able to answer it."

"Action should be taken to review the procedures used in the construction of highways. Silt and dirt pour into the nearest streams from runoff in construction areas."

"Let's have a statewide bond issue to fund parks and outdoor recreation."

"Virginia's air pollution laws and regulations are deplorable. We need more people to enforce the laws and stronger penalties for violators."

"Motor vehicles' exhaust systems ought to be checked at the same time as the auto inspection."

"Studies should be conducted on the extent of groundwater pollution from septic systems."

"Bar the use and sale of high phosphate detergents."

"The Governor should act against the philosophy of water quality control that advocates the addition of more water to bring untreated waste, or partially treated waste, up to some arbitrary standard of dilution."

"Virginia should take action to stop the construction of the Salem Church Dam."

"Stronger legislation is needed to correct the deplorable strip-mining situation in southwest Virginia; for example, require a reclamation plan before a permit is issued."

"Virginia is the only Atlantic Coast state without laws to protect its wetlands."

"The state should ban outdoor billboards."

"Virginia needs to adopt and implement flood plain zoning."

We made front-page news each morning after the hearings. The enthusiasm of the first Earth Day less than a year earlier was translated into specific actions people wanted to see happen. There was no doubt about the public's demands for action by the state government and its environmental agencies. The Council—and the governor—now had an agenda—and I had a lengthy to-do list that would bring it to life. We published a summary of the hearings soon thereafter, entitled "Our Commonwealth . . . Virginians View Their Environment," which became a reference document for establishing priorities.

Later that year, the Council also issued the first-ever report on the condition of Virginia's environment.[1] Among the major topics covered were the following:

Water pollution: Combined sewer overflows are a costly problem to fix. Industrial pollution discharges threaten Virginia waters with a variety of harmful substances, including chemicals, heavy metals, and oxygen-demanding wastes. This problem was emphatically called to my attention by Newton Ancarrow, an early and outspoken advocate for restoring the James River,[2] which had become polluted by municipal and industrial waste in the twentieth century. In 1969 he formed a group called Reclaim the James Inc. When asked about specific polluters of the James, Ancarrow replied that the State Water Control Board lists about 165 industries in the area that were contributing to the pollution. The combined sewer overflow problem, a prominent source of pollution of his section of the James River at the time, remains a major pollution source to this day. In the early 1970s, Ancarrow went to federal district court to try to stop the Corps of Engineers from destroying waterfront vegetation and wildflowers by littering the area with large amounts of debris. He lost the case, but the judge was none other than Robert R. Merhige Jr., who told Mr. Ancarrow, "Mr. Ancarrow, you made us conscious of things we should have been conscious of before."

Erosion and sedimentation: As we explained in the report, erosion and sedimentation had "become a major problem in Virginia accelerated by man's activities." The idea of nonpoint sources of runoff from farms was barely mentioned, though in hindsight perhaps it should have been. However, the report cited the growing problem of urban development, particularly highway development, as causing fifty times more erosion than farmland, and it called for the highway department to reduce erosion on its projects.

Air pollution: In the early 1970s, air pollution from power plant emissions, leaded gasoline, mass transit needs, and automobile emissions was a serious problem with little regulation to combat it.

Groundwater: The loss of groundwater was a major concern. We described in the report how "groundwater withdrawals in certain areas exceed the rate of natural replenishment and no authority exists to regulate this withdrawal."

Wetlands: The wetlands of Virginia were greatly endangered. "Approximately 90% of Virginia's 332,000 acres of wetlands are privately owned. Each year hundreds of acres are drained, dredged, filled, or built on for commercial or other purposes. The state must preserve and protect its wetlands for future generations, but no clear-cut authority exists to do so." The following year the General Assembly enacted the Tidal Wetlands Act, a good start but hardly sufficient to the need.

Other problems such as growing amounts of solid waste, littering, abandoned automobiles, noise, pesticides, and billboards also received attention in the report.

One section was especially prescient: "Land use and development is a fundamental determinant of environmental quality and existing mechanisms and policies for land management are inadequate to carry out the constitutional charge in Article XI." This theme would reappear in the Virginia Environmental Endowment's priorities years later and help lead to the passage of the Chesapeake Bay Preservation Act of 1988, which made official the connection between land use and water quality.

More than fifty years later, there is still work to be done to address these issues. In my view, doing a better job of controlling how and where people build might be more effective than depending on the after-the-fact manipulations and complexities of the TMDL strategy of the Clean Water Act. In other words, wouldn't it be better to limit what's going into the waters in the first place? The same principle applies for eliminating point source discharges.

During his term, Governor Holton initiated and signed into law a wetlands protection act, a soil erosion control law, and an environmental impact reporting requirement for state agencies contemplating capital projects, and led the investment of hundreds of millions of dollars into upgrading older sewage treatment plants and building new ones. He also initiated the development of the first elementary school curriculum for teaching about the environment. Each of these efforts foreshadowed the improvements to Virginia's law, public policy, and environmental regulation that the Virginia Environmental Endowment would later champion and that so many of its grantees would work to achieve.

Governor Holton sought to modernize the state government. He created the state cabinet to better manage the sprawling bureaucracy (about 125 state agencies reported directly to the governor at the time). He modernized agencies who desperately needed more automation of their services, such as the Tax Department and the Division of Motor Vehicles. On top of all this, he embarked on a progressive pollution control and conservation agenda.

The 1972 session of the General Assembly resulted in some useful new environmental laws championed by the Governor. The Virginia Environmental Quality Act was adopted. It declared the state's policy to promote the wise use of the Commonwealth's air, water, land, and other natural resources and to protect them from pollution, impairment, or destruction. It also established in law the Council on the Environment to implement this environmental policy, including the "initiation, implementation, improvement, and coordination of environmental plans, programs, and functions of the State." John W. Hanes Jr. became chair. The Council's scope of responsibility extended to all state agencies and included the specific mandate for assuring that all the existing and proposed public policies of the state be consistent with environmental policy—an ambitious aspiration, if ever there was one, but also consistent with the constitution.

The Tidal Wetlands Act of 1972 declared the state's policy to protect its coastal marshes and wetlands. This was an important step along the path to acknowledging the need to protect coastal areas as valuable natural resources. This bill was championed by a state senator from Northern Virginia in his first term named Joseph V. Gartlan Jr. When Joe joined the Senate in 1972, it was still dominated by traditional Byrd Democrat conservatives. To focus on how he differed from the rest of the Senate would be to miss the point, however; in the following years, he mastered its intricacies, made friends across the dividing lines, and helped to transform an institution known as the burying ground for legislation into a place where legislative accomplishments accumulated.

He was a lawyer who lived in Fairfax with a law practice in DC. He was an eloquent orator, a compelling storyteller, and a passionate advocate for the environment, the mentally and physically disabled, and the poor. He enjoyed it all and was never as happy as when he won a vote on something that mattered. Often, he had the votes because he never forgot that while a good case is necessary, having the votes is what counts.

Over the years, he would occasionally pull me aside and say, "Hey, have you heard this one?" He had an endless supply of Irish jokes and could tell them with the timing of a late-night comedian—he was that good!

He understood the power of stories and he used it to persuade his colleagues to vote for bills that might otherwise not have been supported. He could make the plight of the Chesapeake Bay, the mentally disabled, or the poor more *real*, portraying them as people who were counting on the Commonwealth to do the right thing. Right action was his passion.

He was a champion for the environment and the Chesapeake Bay long before it was politically popular. In 1978, Senator Gartlan cochaired the bistate Chesapeake Bay Legislative Advisory Commission, together with his Maryland counterpart. To further efforts by all the neighboring states to restore the ailing Chesapeake Bay, he introduced legislation creating the Chesapeake Bay Commission in 1980. He was elected as the first chairman of the Bay Commission, served again as chair in 1983 and 1985, and remained one of the body's most dedicated and hardworking members for nineteen years.

We would not have gotten the wetlands protection bill out of committee without Joe's help. Countless acres of wetlands have been preserved because of the precedent that bill set. It foreshadowed what in 1988 became the Chesapeake Bay Preservation Act. Senator Gartlan, also the Senate patron of the Chesapeake Bay Preservation Act, served as a champion for the environment until his retirement from the Senate in 2000 after twenty-eight years.

State Senator (later Congressman) Herb Bateman initially argued in opposition to parts of the wetlands bill, stating that its emphasis on natural boundaries conflicted with property boundaries and property rights, an argument that persists to this day. Nature, however, does not care about property lines, and we need to find ways to accommodate that. This is particularly crucial now that the effects of climate change are causing local officials to respond to rising sea levels, whose consequences for property rights are as much a concern as the environmental damage they inflict. In the end, the bill succeeded and became the Tidal Wetlands Act, and it recognized in a compromise that property lines had to be respected even if nature did not.

The State Water Control Board water quality responsibilities were consolidated with the state's water resource management functions, and the legislature appropriated many millions of dollars for the state to match federal funds for reducing water pollution by constructing new sewage treatment plants and expanding old ones. It also tripled the Air Pollution Control Board's budget and significantly increased funding for the acquisition of state parks.

In 1973 the Council tried to consolidate the various environmental agencies into one department, an environmental protection agency that would combine the air, water, and waste agencies. We hired a young woman named Elizabeth Haskell to assist us with this project. Haskell had been a scholar at the Princeton Institute for Advanced Study and had just written a book about recent environmental agency consolidations in several other states. She was an enthusiastic and knowledgeable expert on such matters. Not surprisingly, the agencies we proposed to consolidate unanimously opposed the idea, and the General Assembly wasn't having any of it.

In 1973 Governor Holton named Elizabeth Haskell to serve as a member of the State Air Pollution Control Board, where she eventually became chair. Sixteen years later, she became the first woman to serve as Secretary of Natural Resources, when Governor L. Douglas Wilder appointed her in 1990. And, as an example of how "coming events cast their shadows before them," one of the most important accomplishments of her term was to persuade the governor and the General Assembly to create a new Department of Environmental Quality, fulfilling at last her plans from the Holton era.

Also in the 1973 session, Delegate James H. Dillard II successfully guided House Bill 766, the Erosion and Sediment Control Act, to passage. This law was the first policy recognition of the relationship between the use of land and the quality of water and was a forerunner of the Chesapeake Bay Preservation Act in 1988.

Also in 1973 we thought it might be useful to have a state law similar to the federal National Environmental Policy Act (NEPA), because construction of new state agency projects, such as a new prison, for example, did not have to consider the impact of such projects on the environment at all. We based our draft on the mandate for environmental analysis in Article XI and modeled our proposal on the federal National Environmental Policy Act.

NEPA, still a new idea, was sponsored by Senator Henry Jackson and Congressman John Dingell. It passed the Congress with overwhelming bipartisan support, and President Nixon signed it into law on January 1, 1970. NEPA established a national policy for the environment and provided for the establishment of a Council on Environmental Quality within the Office of the President: "NEPA sets forth a national policy to use all practicable means and measures, including financial and technical assistance, in a manner calculated to foster and promote the general welfare, to create and maintain conditions under which man and nature can exist in productive harmony, and fulfill the social, economic, and other requirements of present and future generations of Americans. *42 U.S.C. 4331(a)*."[3]

NEPA was a remarkable piece of legislation, both in its scope and in its brevity. It covered all decisions made by the federal government and had at its core the truth of old wisdom: "measure twice and cut once." It required all federal agencies to evaluate the environmental impact of their decisions *before* making them; to think through what and how they proposed to do something; to analyze alternative approaches, including a "take no action" alternative; and to document that analytical process clearly and completely for public review. It also established the "environmental impact statement" to report the details of those decisions and gave the Council on Environmental Quality the responsibility to oversee the implementation of NEPA. What a simple, brilliant idea.

Many of those federal decisions, of course, are sought by and carried out by private corporations seeking federal contracts to build roads, bridges, pipelines, and other large infrastructure projects. As such, they too are required to think through and analyze the alternative environmental impacts of their proposed projects and present their analysis to the relevant federal decision-making agency.

The Environmental Impact Report law we intended would require a state agency proposing a capital expenditure of more than $100,000 ($667,000 today) to prepare a report explaining the project's impact on the environment and the agency's plans for minimizing that impact. The law was precipitated by the battle over the Department of Corrections' plan to build a prison in the Green Springs neighborhood near Charlottesville, a battle they lost but not without leaving it apparent that Virginia needed a law that would require agencies to think these ideas through before proceeding with them, particularly their potential impact on the environment. The Council on the Environment would have the responsibility to review and comment on these reports and advise the agency and the governor regarding how to minimize each project's environmental impact.

Getting that law approved was a lot of fun, with a fair number of twists and turns in its path through the legislature, particularly the State Senate. We got the bill through the Agriculture and Natural Resources Committee without a problem, but it needed a $30,000 ($200,000 today) appropriation to hire someone to administer the program. That meant the bill had to go to the Finance Committee, headed by the senator from Richmond, Ed Willey.

Fortunately, I had developed a good working relationship with Senator Willey during my three years in the government, and we chatted frequently, usually while walking the corridors of the historic state capitol building, which at that time was where all official General Assembly business was done. A tall, white-haired man who owned a drug store, he was always impeccably dressed in a dark suit, white shirt, and conservative tie. Amidst the daily hustle and bustle of that place, catching up with him was always a challenge. He talked, I listened. He had a reputation as a curmudgeon, but to me he was always friendly and helpful.

Senator Willey served as president pro tempore of the Senate and chaired the Finance Committee. When I asked him to approve the appropriation, he replied that the highway commissioner, Douglas B. Fugate, thought his agency ought to be exempt from the proposal because it was already required to file extensive and expensive reports with the feds under NEPA. We still wanted the Highway Department included at the state level, because many of the state's secondary road projects were not covered by NEPA. However, Senator Willey looked at me, smiled, and said, "Jerry, if you want the bill, drop the Highway Department." So we did, and the bill passed easily, making Virginia one of the first states to have its own "little NEPA."

A further step toward improving water quality came with the enactment of Senate Bill 387, which authorized the State Water Control Board to regulate groundwater in Virginia.

We also persuaded the legislature to direct the State Board of Education to establish a program of environmental education in the public schools of the Commonwealth. By the fall of 1974, the Department of Education delivered the first environmental education model curriculum to every school district in Virginia. It was a modest start, but it later gained momentum with the formation of the Virginia Association of Environmental Educators (VAEE). VAEE included among its members teachers and nature center experts, who continue to bring the concepts of environmental education alive to thousands of children annually.

During the Holton administration, I also made a good friend in Jerry Baliles, who was an assistant attorney general and our Council's first legal advisor. A few years later he was elected to the House of Delegates, then he became Attorney General of Virginia and finally governor in 1985. Like Holton, he, too, distinguished himself as an environmental governor.

While environmental problems persist, such as the effects of climate change, the Holton administration helped to create what Holton often called a "New Dominion," exemplified by economic development, equality for all people, and environmental improvement. A solid foundation of legal and policy infrastructure was put in place to support the environmental progress of recent decades.

# 10

# The Updated Mission of
# the Endowment

THE VIRGINIA ENVIRONMENTAL ENDOWMENT commenced operations in 1977 with an initial capitalization of $8,000,000 as a result of the federal Kepone trial. In 1981 VEE received another $1,000,000 from the FMC Corporation as part of the settlement of a federal plea to the charge of discharging carbon tetrachloride into the Kanawha and Ohio Rivers. The spill moved steadily downstream, contaminating municipal fresh drinking water intakes from Charleston, West Virginia, along the length of the Kanawha River. Its effects extended out of the Kanawha and into the Ohio River from Point Pleasant, West Virginia, all the way downstream to its mouth at Cairo, Illinois, where it meets the Mississippi.

When the Department of Justice and the Environmental Protection Agency settled the FMC case by accepting the $1 million payment, they asked local governments along the rivers for ideas on how to use the money. Meanwhile, Bill Cummings, who had just stepped down as US Attorney, contacted the US Attorney in Philadelphia, whose office had prosecuted the case. Cummings told him how the Justice Department had resolved the Kepone pollution case, and he volunteered to be helpful if needed. A short time later, a lawyer at the EPA contacted me to learn about VEE and what we did, how we operated, and so on. "That makes a lot more sense than some of the proposals we have heard," she said. She went on to give me a few examples of what had been proposed for how local jurisdictions might receive some share of the million dollars—and to say she was not inclined to support any of them. "How would the Endowment feel about taking this project on?" she asked.

Bill Cummings was helpful in setting the criteria under which the Endowment might accept these funds, while I outlined how we would be able to grant them to eligible entities for water quality purposes. In short, a deal was struck. The federal government ordered FMC to send the money to VEE, and we amended our charter to expand our activities and purposes beyond Virginia.

This allowed VEE to expand its water quality grant-making to West Virginia, Kentucky, and Ohio. We called this new program the Kanawha and Ohio River Valley Water Quality Program.

The efforts of the Kanawha and Ohio River Valley Water Quality Program in Kentucky and West Virginia have made a big difference. VEE received $1 million from FMC in late January 1981. The Endowment started out by making a variety of grants aimed at improving water quality. Our first grant for environmental advocacy in Kentucky was made in 1984 to a new group, the Kentucky Resources Council (KRC), a Kentucky-based nonprofit founded by Tom FitzGerald. KRC was both a fledgling effort and a successor to a long-standing program. "Fitz" abhorred pollution and environmental injustice and spent much of his time in the early years helping people in the coal mining region of eastern Kentucky. He had been a staff attorney at the Appalachian Research and Defense Fund of Kentucky (AppalReD), providing free civil legal services on environmental problems to low-income individuals and community groups in eastern Kentucky.

Eventually, the Kanawha and Ohio River Valley Water Quality Program settled on a main focus: helping to start and sustain advocacy groups. Two groups in particular continue as effective advocates and watchdogs today, the Kentucky Watershed Watch, led so effectively by Hank Graddy, and the West Virginia Rivers Coalition, whose director, Angie Rosser, has made it a force for clean water in West Virginia.

The Reagan Administration in 1981 marked a turning point in the consensus politics that had resulted in adoption of major environmental laws in the 1970s. It also affected the ability of grantees of the nonprofit Legal Services Corporation to help low-income individuals and community groups regarding environmental issues. AppalReD was one of those grantees. AppalReD was involved in litigation, education, and lobbying efforts involving the federal 1977 Surface Mining Control and Reclamation Act. The limitations imposed on support for legal services affected the ability of AppalReD to represent clients on environmental matters, which led Tom FitzGerald, a young staff attorney at the time, to leave AppalReD and to reshape the Kentucky Resources Council into a nonprofit provider of pro bono legal assistance to individuals and organizations that could not find or afford representation.

FitzGerald describes the involvement of VEE this way:

When I came to KRC, it was a shell of an organization with tax-exempt status, no staff, and an inactive board. I set about rebuilding it as a non-governmental legal aid firm, providing free legal assistance, accepting no government or corporate funds. Never having written a grant proposal, I wrote and mailed around 40 such proposals. The few that even responded indicated "we don't fund legal work" or suggested that I talk with "grassroots" groups about funding.

Out of those scores of proposals, only VEE and its Director "got it." As a former state environmental official, McCarthy explained that he understood the technical, legal, and other barriers to effective grassroots participation in environmental governance decisions, and took a gamble on an organization with a proud heritage (KRC had evolved from the former Kentucky Rivers Coalition, which was instrumental in opposing the Red River Dam and other Corps of Engineer dam projects in Kentucky during the 1970s) and a Director with a vision but only a brief track record as a staff attorney with a legal aid firm.

In the first and pivotal years of KRC's work, fully half the funding came from VEE, with the other half coming from Mary Bingham, whose family owned the *Louisville Courier-Journal,* and who as a member of the Kentucky Environmental Quality Commission, had observed FitzGerald during meetings where he presented testimony on mining and other issues. It is not hyperbole or overstatement to suggest that without VEE's understanding of the barriers facing those most affected by environmental decisions, and his faith and willingness to take a risk, KRC would not have survived and the history of environmental protection since 1984 in Kentucky would have been very different.[1]

Fitz stepped down as KRC's leader in October 2021 after thirty-seven years of representing the interests of those whose voices are not always heard.[2]

Since its inception in 1989, the West Virginia Rivers Coalition (WVRC) has been working to promote the conservation and restoration of West Virginia's freshwater resources. The West Virginia Rivers Coalition continues to advocate for river protection and enhancement by steadfastly following its original mission "to seek the conservation and restoration of West Virginia's exceptional rivers and streams."[3]

Over a forty-year period, the Endowment's investments in clean water through the Kanawha and Ohio River Valley Water Quality Program had a major impact on water quality in Kentucky and West Virginia. During my tenure, at least $3 million went to fund water quality improvements in the Kanawha and Ohio River Basins.

WHEN IT WAS CREATED, VEE became the only grant-making organization in the country that focused 100 percent on environmental grant-making as its mission. As a way of building relationships with other funders as well as wider support for environmental grant-making, the Endowment joined the Council of Foundations and the Southeastern Council of Foundations (now renamed Philanthropy Southeast), and in 1987 it became a founding member of the Environmental Grantmakers Association. At about the same time it became one of the charter members of the Association of Small Foundations, now called Exponent Philanthropy. In the 1990s, we were one of several foundations that started the Chesapeake Bay Funders Network (CBFN), which expanded grants for environmental purposes throughout Virginia and the Chesapeake Bay region.

Initially, VEE focused on scientific and public policy research in an effort to reduce toxic discharges into state waters and, we hoped, reduce the likelihood of another Kepone-scale incident. The Endowment focused its grant-making on water quality topics and translating national environmental concerns into local action in Virginia, because while environmental concerns are global in scope, they are local in their effects on people's lives. In its first five years the Endowment contributed $2.4 million to seventy-five projects that, with matching funds, totaled $5 million in value. We clarified our mission to focus on bringing people together to protect water quality, conserve natural resources, and promote environmental literacy. We also published an annual report each year so that people interested in applying for grants could see what our priorities were and what kinds of grants we were making. Compiling the annual report also served as a way for us to review our program goals, annual objectives and financial condition and to make any necessary changes to our work plan.

Among the major accomplishments of the first few years was the establishment of the Institute for Environmental Negotiation at the University of Virginia. We also made the first grants to strengthen citizen advocacy efforts, which were conducted at that time mostly by volunteers. We helped the University of Richmond Law School purchase and maintain a comprehensive environmental law library. We strengthened the environmental law program at the College of William and Mary's Law School and provided funds to the University of Virginia Law School to create a new law journal, the *Virginia Journal of Natural Resources Law,* subsequently renamed the *Virginia Environmental Law Journal.* The Endowment also funded the establishment of a new doctoral program in Environmental Biology and Public Policy at George Mason University.

VEE grants enabled the Chesapeake Bay Foundation to open its Virginia Office in Richmond in 1980, a great boon for Bay protection efforts in Virginia ever since. Beginning also at that time, grants to the Bay Foundation helped to establish an environmental education program. Will Baker, longtime President of CBF, estimated that "at least a hundred thousand students have benefited from this over time. Our hope is things will be better for future generations as they become adults and say, we've got to take care of this planet."

In 1993 the Endowment also made a $75,000 grant to CBF to establish the Admiral Ross P. Bullard Environmental Education Scholarship Fund for youth in the Hampton Roads area. CBF's environmental education programs subsequently secured state funds for continued support of its excellent programs.[4]

CBF was graced by two remarkable educators, John Page Williams and Bill Portlock. At different times I had the pleasure of their company while they took me on field trips to see Bay tributaries "up close and personal." John Page and I once drifted along Dragon Run on one of those tiny boats, which I was sure was going to sink at any moment, but John Page's nonstop exposition of the details of the plants, animals, and fishes that we were seeing kept me from worrying much about that. He could really tell a story! And he knew about every bit of nature we encountered on that hot and muggy late spring day. VEE made a grant to the Friends of Dragon Run not long after their formation in 1985.[5]

Bill Portlock introduced me to eagles around the same time. We went for a cruise on the Rappahannock River in a small boat that could hold maybe a dozen people. We were looking for eagles, which were few and far between at the time. We met in Tappahannock, boarded the boat, and headed north toward Fones Cliffs, one of the most pristine locations on the East Coast to view bald eagles. We were overjoyed to see a couple of the great birds. More common were ospreys, but the eagles were the stars of that beautiful sunny day as they slowly swooped above and around the cliffs as we watched mesmerized by the rare sight.

Today, Fones Cliffs remains pristine thanks to the efforts of The Conservation Fund and its partners. The result is that more than a thousand acres of Fones Cliffs, including almost three miles of cliff shorelines, have been officially protected and will remain unhindered by development along with the adjacent Rappahannock River Valley National Wildlife Refuge and Rappahannock tribal lands. The Rappahannock River is a major destination for seeing eagles today.

Efforts to improve the local environment got another major boost when the Chesapeake Bay Mini Grant Program, which had been created by the Citizens Program for the Chesapeake Bay, benefited from initial funding from VEE, among other sources. Inspired by the spirit of volunteerism, the program has carried out a lot of local environmental projects without a lot of money. After three years this mini-grant program was extended statewide with grants we made to the Virginia Humanities Council.[6]

Cooperation between the worlds of business and environment has been successfully demonstrated in the establishment of a mainland headquarters by The Nature Conservancy on the Eastern Shore. Another program, intended to establish a waste exchange in cooperation with the Virginia State Chamber of Commerce, failed to attract enough businesses to make it work.

Starting in 1977 with $8 million for Virginia programs and adding $1 million in 1981 for the Kanawha and Ohio Rivers Water Quality Program, the Endowment in its first ten years made 208 grants totaling $5.9 million and still had $18 million to continue its work. Many of the initiatives VEE helped to launch during this period, such as The Nature Conservancy's Virginia Coast Reserve, the James River Association, the Southern Environmental Law Center, the Institute for Environmental Negotiation, and the Kentucky Resources Council, continue to improve the quality of the environment with more effectiveness than ever before.

# 11

# The Public Interest

THE 1970s saw major environmental laws passed with overwhelming bi-partisan support. When the Endowment began its work, environmental protection and natural resources conservation were consensus national priorities with states playing a supporting role. That started to change in 1981 when President Reagan's administration placed unprecedented ex-pectations on state governments to develop, manage, and fund their own environmental programs in ways that best fit each state's circumstances.

When we started to fund projects addressing toxic substances in Vir-ginia's rivers and streams, we were concerned that the existing laws were not being enforced. We commissioned a study of the effectiveness of the water permit program regulating discharges into the James River. Thanks to the research conducted by the National Wildlife Federation, Virginians gained an idea of what was being discharged into the James River. The re-sults of that analysis showed that the state permit records and the permits were years out of date and that discharges were occurring that the state did not even know about.

We adjusted our priorities to try to strengthen environmental permit-ting and law enforcement. During that time, citizens had no professional advocates representing them before the state regulatory boards that regu-larly dispensed air, waste, and water pollution control permits. The par-ticipants in such deliberations usually only included the regulators and the regulated—that is, the only two entities involved were the organization requesting the permit and the state, personified by the regulatory board and advised by the attorney general's office to ensure that the law's proce-dural requirements were satisfied.

We saw an opportunity here to make a difference and set out to build a professional cadre of advocates for the public interest. We needed to get a public interest law firm involved to represent the public's interest before these regulatory boards.

The negotiation of a permit at that time was limited to what the permit-tee wanted and what the agency was willing to permit; the rest of us, who must live with the consequences of all that toxic material pouring into the rivers, were not a part of it.

I noted that participating in regulatory proceedings requires a level of technical and legal knowledge the ordinary citizen does not possess. Nor, in 1980, did we have such expertise available in Virginia.

The Endowment had recently offered to fund the Chesapeake Bay Foundation to open a new Virginia office, but it was just getting started. It did not have a full-time lawyer participating regularly in state air and water regulatory actions. We had also funded the Conservation Council of Virginia's executive director position. But both were focused on the legislature, specifically on lobbying for good environmental laws. No one was doing regulatory work.

From the Endowment's point of view it seemed only fair that in such regulatory matters the private interests seeking a permit ought to be balanced against the public interest seeking to protect environmental quality. We decided to look at national environmental law groups to see if someone could help.

The Environmental Defense Fund (EDF), based in New York City, had recently opened a Colorado office to focus on that state's environmental problems. EDF's director, Fred Krupp, was open to the idea of setting up a Virginia office if VEE could provide the major funding of it for a couple of years. We were, and we did.

EDF wisely decided to hire Assistant Attorney General Timothy G. Hayes to direct its new Virginia office, and he went right to work. Having previously represented the State Water Control Board in permit hearings and proceedings, he knew exactly where and how to apply his knowledge and pressure to have regulatory decisions be more attentive to their effects on water quality. He did a great job representing the public interest before the regulatory boards.

After a few years of outstanding work in the public interest, Tim left EDF for private practice, having strengthened the permitting process to account for protecting the air, land, and water from pollution, impairment, or destruction. He was also a member of the Toxics Roundtable, an informal group of environmental advocates and chemical industry leaders who were negotiating common ground on regulating toxic discharges. Their principal accomplishment, with the assistance of the Institute for Environmental Negotiation in a process it called a "policy dialogue," was to draft a comprehensive Hazardous Waste Facility Siting law, which Virginia enacted in 1986 to guide the siting of new hazardous waste disposal facilities.

EDF kept the office open for a couple more years, choosing as Tim Hayes's replacement David Bailey, a lawyer who had previously served as an engineer at the State Water Control Board and who continued the good work EDF was doing. However, raising funds was a constant problem, and soon EDF decided to change strategy and move away from operating state offices to focusing on national and international issues. We were disappointed that EDF decided to close its Virginia office. After five good years of having the public interest represented in environmental regulatory proceedings, we no longer had that capacity in Virginia.

Meanwhile, the regulatory agencies continued to issue permits for all manner of waste discharges. To this day, there remains a staff at the Virginia Department of Environmental Quality whose principal function is to issue permits.

And then we got lucky. One morning in 1986 I got a phone call from my friend Pat McSweeney. Pat is a lawyer I had known since 1971 when he returned to Richmond to serve as executive director of the Hopkins Commission, a legislative study commission that, in concert with the Holton administration, recommended a major reform of state government: the establishment of a governor's cabinet to better manage the more than one hundred state agencies that existed at the time.

"Jerry! How are you? I've got a guy you need to meet. His name is Rick Middleton, and he is just setting up a new environmental nonprofit law firm in Virginia." When one door closes, another opens? My hopes rose immediately.

Soon after, both Pat and Rick came to my office to tell me what the new group hoped to do. Here's how Rick Middleton tells the story.

I met Pat when I was in Washington, DC, and he was working on some case and wanted some expert advice. So, he came up to DC just to hash the case over with me. I knew him sort of, we were in the same fraternity, but Pat had graduated before I got there. When I was at the University of Virginia, his law partner was James Stutts.

And James I knew very well. You know, we were friends at UVA. So I looked at James and Pat as being one of my major entry points here in Virginia, and to introduce me to Jerry McCarthy and the Virginia Environmental Endowment. I learned all about it through Pat. He set up a meeting, we came to your office, and he introduced us, and I told the SELC story to you.[1]

From the Endowment's point of view, the arrival of the Southern Environmental Law Center (SELC) on the scene was attractive and timely, because we needed help to get the public represented in regulatory proceedings. As EDF was closing its doors, there was no longer any environmental law expertise readily available. Most of the groups that we take for granted today didn't even exist. Rick continues the story:

> Well, that was incredibly lucky for me, quite possibly because SELC wouldn't have made it without VEE's timely support. I remember when I met Tim Hayes, he had already just stepped down. He was supportive. I didn't want to step on anybody's toes, and I really didn't want to compete with anybody. I'm not even sure Bud Watson was still at CBF in 1986. He was a lawyer, but their focus wasn't really doing legal work.

Rick Middleton, a native of Birmingham, Alabama, graduated from the University of Virginia and earned his law degree from Yale. Starting SELC with just himself and another lawyer, David Carr, he established this nonprofit law firm in 1986, and VEE became one of its first supporters. Soon after, lawyers Kay Slaughter and Deborah Murray joined SELC. The mission of the Southern Environmental Law Center remains the same today: to use the power of the law to protect the environment of Virginia and the Southeast as well.

The power of the law has been a force in environmental protection since the early 1970s, but nowhere has it been put to more effect than by SELC in Virginia and many other southern states. SELC has compiled a remarkable record of achievement since 1986, using policy research, advocacy, persuasion, and when necessary, litigation to improve and defend environmental quality. Headquartered in Charlottesville, it now operates in nine southern states to present alternatives and constructive solutions for preventing damage to natural resources, including pollution of water, wetlands, and forests. More than eighty lawyers now work daily to carry out SELC's mission.

SELC works with all three branches of Virginia's government to shape and enforce the laws and policies that determine the quality of air, water, and landscapes as well as related issues of land use, transportation, and community quality of life. In keeping with the Endowment's original decision to avoid supporting litigation, VEE and SELC agreed that VEE grants would only be used to support legal research, policy development, and advocacy before legislative and regulatory bodies in Virginia. Since 1986, SELC has made an impact on water quality, wetlands, forests, land use, and transportation policies. It has represented other groups in litigation to fight natural gas pipelines, clean up coal ash pits, and defeat a major reservoir project that would have resulted in the largest authorized destruction of wetlands in the mid-Atlantic since Congress passed the Clean Water Act in 1972.[2]

One of the most significant of SELC's many accomplishments on behalf of a clean environment occurred when SELC attorneys Jeff Gleason, Blan Holman, and Cale Jaffe won a major Clean Air Act case before the Supreme Court in 2007.[3] The case was *Environmental Defense v. Duke Energy.* SELC represented the Environmental Defense Fund, which at that time was known by the shorter name Environmental Defense. Quoting from the summary of the case: "In the 1970s, Congress added two air pollution control schemes to the Clean Air Act: New Source Performance Standards (NSPS) and Prevention of Significant Deterioration (PSD), each of them covering modified, as well as new, stationary sources of air pollution."

In a unanimous opinion by Justice David Souter, the Court ruled that the EPA need not interpret "modification" in regulations designed for the "prevention of significant deterioration" the same way the term is interpreted in newer regulations of source performance standards, that is, as the level of pollution a new source of discharges may produce. The Court held that the "EPA's construction need do no more than fall within the limits of what is reasonable, as set by the Act's common definition." The case ensured a multibillion-dollar cleanup of the oldest and dirtiest coal-fired power plants in the Southeast and helped lead to the retirement of some of those coal units, reducing millions of tons of greenhouse gas pollution. It was a unanimous victory in defense of the Clean Air Act.

Protecting special places is also part of SELC's portfolio. In March 2009 SELC won passage of the Virginia Wilderness Law. With a 285–140 vote in the US House, Congress passed the bill on March 25, permanently protecting more than 53,000 acres in the publicly owned Jefferson National Forest.

Stopping the King William reservoir proposal was a significant victory for clean water and SELC. Deborah Murray, the SELC lawyer who represented several opponents, commented:

> To me, the defeat of the proposed King William Reservoir remains the most meaningful and significant achievement that I have played a part in over my many years at SELC. The reservoir would have destroyed more than 400 acres of wetlands—what would have been the largest single "authorized"

destruction of wetlands in the mid-Atlantic since the Clean Water Act was enacted. Newport News planned to locate the intake structure for water withdrawals from the Mattaponi River in prime shad spawning habitat, impacting shad, and threatening the cultural heritage and traditional way of life of the Native Americans. The Norfolk District of the Corps of Engineers, to its credit, ordered its own independent studies of Newport News' projections and reached the conclusion that the reservoir was unnecessary: Newport News had greatly inflated its future water needs and the region's legitimate needs could be met without the reservoir. In 2001, in a 300+ page analysis, the Norfolk District recommended denial of the § 404 permit because of the severe environmental impacts, the impacts on Native Americans, and the lack of need for the project. The final decision was up to the North Atlantic Division of the Corps. The North Atlantic Division brushed aside the Norfolk District's comprehensive analysis and, in 2005, issued a § 404 permit for the project.

She added, "We filed suit in 2006, and the U.S. District Court ruled in our favor in March 2009 and invalidated the § 404 permit; not long thereafter, Newport News cancelled the project. The decision was not appealed."[4]

As late as the 1990s, citizens still had to persuade state courts that they had the right to be heard in permitting decisions. One of the most important rulings on this matter came about because of SELC's efforts. SELC played a significant role, together with several other environmental organizations, including the Chesapeake Bay Foundation, the Sierra Club, and the Environmental Defense Fund, in pushing for the right of citizens to be able to challenge permitting decisions made under the Commonwealth's water, air, and waste laws. Deborah Murray told me:

As was apparent in cases such as the decision in *Town of Fries v. State Water Control Board*, 13 Va. App. 213, 409 S.E.2d 634 (1991), Virginia courts interpreted the term "aggrieved" in the judicial review statute to bar anyone—other than a disappointed permittee or permit applicant—from challenging an agency permit decision. Our organizations repeatedly raised the issue of the lack of standing in the courts, in amicus briefs, in communications with EPA, and with sympathetic Virginia lawmakers, building the case to change the statute.

The matter came to a head when Virginia sought EPA's permission to administer the Title V air permitting program. Because Virginia law did not provide standing equivalent to that of Article III of the U.S.

Constitution to challenge agency permitting decisions, as required under the Clean Air Act, EPA denied the application. The Commonwealth sued EPA, and our organizations intervened in the lawsuit. In 1996 the Fourth Circuit Court of Appeals ruled in EPA's favor, thus forcing Virginia to change its judicial review statute to obtain permission to administer the Title V program. The General Assembly, reflecting the growing public support for citizen standing more generally, changed the law to provide Article III standing to challenge state permitting decisions issued not only under the air law but also under the water and waste laws.[5]

One more example of the power of the law must include the largest cleanup of toxic waste in the Southeast, which was because of SELC's coal ash litigation, led by South Carolina attorney Frank Holleman. For decades power companies had stored coal ash from power plants on the banks of nearby rivers. The storage systems were unstable and prone to leak into the rivers and groundwater. As described by SELC attorney Deborah Murray, they contained millions of tons of toxic wastes; they were unlined and could leak arsenic, mercury, selenium, and other contaminants. SELC fought in federal and state courts to get utilities to clean up these threats to water quality and public health. In 2015, SELC filed a Clean Water Act citizen suit against Dominion (the region's power company) for the continual long-term discharge of arsenic from Dominion's Chesapeake plant into the Elizabeth River. SELC won at the district court, but the ruling was overturned in the Fourth Circuit Court of Appeals. SELC's Murray stated, "Together with our partners, including Sierra Club, Potomac Riverkeeper Network, the James River Association, Virginia Conservation Network, and the Virginia League of Conservation Voters, we refocused our efforts on the General Assembly. In late February 2019, both chambers, with overwhelming bipartisan support, passed the landmark legislation requiring Dominion to excavate and safely dispose of its coal ash."

As a result of our work, all unlined coal ash pits in North Carolina, South Carolina, and Virginia are being cleaned up and utilities in the Southeast have been required to remove over 270 million tons of coal ash from unlined, leaking pits across the region.

Following our efforts to clean up coal ash contamination leaking from three different Dominion Virginia Power facilities, Virginia enacted legislation in 2019 requiring Dominion to excavate all its coal ash pits in the

Commonwealth. In 2019, the Virginia General Assembly passed legislation that requires Dominion Energy to excavate roughly 28 million tons of coal ash stored in unlined ponds at Dominion's four power plants located in the Chesapeake Bay watershed—the Chesapeake Energy Center, Chesterfield Power Station, Bremo Power Station, and Possum Point Power Station. Under the law, Dominion must recycle a certain percentage of the coal ash into products such as concrete, and must dispose of any remaining ash in modern, lined landfills. This is the first law in the country to require what is referred to as "clean closure" of coal ash ponds.

In April 2020 Virginia Governor Ralph Northam signed the Virginia Clean Economy Act, which will replace all carbon-emitting power plants in Virginia, achieving zero carbon generation by 2050. This legislation will accelerate retirement of fossil-fuel power plants, establish a statewide energy efficiency standard, speed installation of rooftop solar for homeowners, and expand larger-scale solar and offshore wind. SELC worked with the Virginia Conservation Network, the Northam administration, legislators, and many environmental partners to pass this law. The fate of the Virginia Clean Economy Act under the new administration elected in 2021 remains unclear.

One area of policy reform in which SELC has been particularly successful is its Land and Community Program, which VEE has supported for many years. As stated by Trip Pollard, the director of that program: "VEE's investment in the (program) has supported our efforts to educate decision-makers and the public about growth issues, as well as links among transportation, land use, energy, and environmental quality."[6]

SELC's "Beyond Asphalt" report in 1999 provided the first comprehensive examination of Virginia's transportation trends, impacts, and policies.[7] Many of its recommendations for reform were adopted by the Commonwealth of Virginia. I had a front-row seat to much of that reform during my nine years as a member of the Commonwealth Transportation Board (2002–11). Governors Mark Warner and Tim Kaine both embraced transportation reforms and invested in rail and transit programs as well.

The Endowment's long-term commitment to SELC has allowed them to challenge problems that were decades in the making and in which the public, until recent times, has rarely had opportunities to participate in preventing or solving. Transportation is a good example. Just as air, water, and waste boards were once a closed circuit between permission-seekers and decision-makers, many transportation decisions were made in a tight circle of interest as well, principally between the state and the localities around the state who were seeking money to fund their highway and bridge projects. Even the membership of the Transportation Board normally included developers, their lawyers, and others with a stake in CTB decisions. SELC helped change all that, bringing expertise, ideas, and negotiation skills to represent the wider public interest in transportation policies and projects.

The link between land use planning and transportation planning became more obvious and therefore harder to ignore. Among the new developments were state-supported passenger rail service from Lynchburg to Charlottesville and Washington, DC, and new funding for passenger rail between Norfolk and Richmond for the first time in decades. Another innovation was the creation of a passenger rail fund for both capital and operating expenses, the first fund of its kind in the country. As a result, Virginia has in recent years, and with an infusion of federal dollars, invested not only in roads but also in high-speed rail, commuter rail, and local transit systems.

Trip Pollard referred me to US District Judge Terrence Boyle, who stated in one of SELC's cases, "No human activity has a greater impact on where and how we develop and how we impact the environment than the decisions of where to build roads." SELC has successfully challenged dozens of ill-conceived road projects to keep valuable landscapes intact, protect air quality, and better coordinate transportation and land use decisions. In recent years there has been progress to advance the "quiet revolution in land use" first described by Fred Bosselman and David Callies in their 1972 report *The Quiet Revolution in Land Use Control,* which was prepared for President Nixon's Council on Environmental Quality. It has been accomplished in Virginia in an unusual, unexpected, and unprecedented way: by use of the state's transportation system. Since 2006, Virginia has developed a system for coordinating state transportation planning and local land use decision-making and in the process has done more to assert the state's legitimate role in land use planning than almost anything else it has tried over the decades since the Bosselman and Callies report. The state of Virginia accomplished this by a skillful combination of "carrot and stick" involving road fund investment policies and congestion-reduction strategies. One of the biggest challenges facing transportation planners is continued growth in population and development in Virginia and, as a result, the need to make better land use decisions. Improving the coordination between transportation and land use would appear to be an obvious requirement. One key step in that direction was the development of traffic impact analysis requirements mandating that all developments with a substantial impact on the state highway network use the Virginia Department of Transportation's statewide, uniform standards to analyze the impacts of the development on the transportation network.

SELC was an important voice for the public interest in all this activity, and not just in opposing damaging projects like the ill-considered replacement of Route 460 with an interstate-type toll road, or the wholesale rebuilding and widening of all 325 miles of I-81. SELC always offers positive alternatives, because the needs are real but the solutions are often full of alternative possibilities, and for the largest proposals, often less expensive alternatives too.[8]

# 12

# The Chesapeake Bay

AN IMPORTANT part of VEE's strategy is the use of science to advance public policy on the environment. Efforts the Endowment funded to protect and conserve the Chesapeake Bay, one of America's great treasures, illustrate this aspect of its work. In the early 1980s, it was being systematically polluted, overfished, and generally discounted as both an economic and natural resource serving millions of people in Virginia, Maryland, Pennsylvania, and the District of Columbia. Shamed by a newly formed advocacy group, the Chesapeake Bay Foundation, and goaded into specific actions by a tireless citizens advisory committee, the politicians of the Bay region finally agreed to a series of binding agreements that laid out goals, objectives, milestones, and timetables to clean up the Chesapeake Bay.

The Chesapeake Bay Foundation, an Annapolis, Maryland–based nonprofit organization, has become the leading advocate for cleaning up pollution and restoring the natural resources of the Chesapeake Bay. CBF documents the Bay's condition in a biannual report, which is reported on by newspapers throughout the Bay region. It was founded in 1967 by a group of Baltimore businesspeople. They enjoyed regular fishing outings in the Bay and over time became concerned about the pollution of the Bay. In the decades since, the Foundation's efforts to build public support to "Save the Bay" and hold governing bodies accountable for doing so have been the principal driver for improving the quality of the Chesapeake Bay as both a natural and an economic resource.

In 1979 when I first contacted the Bay Foundation, its only office was its headquarters in an old, repurposed church in Annapolis. At the time there wasn't much focus in Virginia on protecting the Bay, although in 1980 State Senator Joe Gartlan introduced legislation creating the Chesapeake Bay Commission, after which he was elected as the Commission's first chair.

VEE wondered if CBF would be interested in expanding its work in Virginia. I met with David McGrath, who had just succeeded CBF founding executive director Arthur Sherwood, and heard his pitch. CBF needed our support for its venture on Smith Island, which lies approximately ten miles west of Crisfield, Maryland, across the Tangier Sound portion of the

Chesapeake Bay. The first project involved taking small groups of students to Smith Island for real-life exposure to the Bay, its resources, and its value to the environment. The second part of the request was to support a national estuarine study center there.

My office was barely large enough for two people to sit comfortably, and its appearance was simple, with secondhand furniture accented by a couple of John Barber prints hanging on the wall. David had a nice way about him, and I could tell that while he was new to fundraising conversations, he was an effective advocate for his cause. It's always a good sign when someone starts waving their arms to punctuate a point, and he did that. I didn't mention the Virginia office idea, though. This was the first time we had considered a grant to CBF, and we didn't know the group well yet. Looking back on it now, one wonders how we could have ever doubted that CBF would be a huge success!

We made a $14,000 grant plus a $14,000 loan for the Smith Island project. Organized philanthropy calls this a "program-related investment," a technique for providing support, sometimes used in lieu of an outright grant when a foundation might want to use some of its assets rather than its grant funds to accomplish a worthy purpose. Program-related investments were a sparkling new tool in funders' toolkits at the time. It lowered our risk a bit to use this approach, in that we could always ask for the loan to be repaid. CBF accepted the arrangement.

In 1980, David McGrath visited the VEE office to discuss progress on their work. He asked if we might convert the $14,000 loan into a grant to continue the Smith Island work or, alternatively, increase the Foundation's activities in Virginia. Since CBF was the only citizen-based advocate for saving the Bay, he suggested the establishment of a Virginia office, which could be useful in building support in Virginia for CBF's efforts. He had been thinking along the same lines as VEE regarding the Virginia office idea, and I was delighted to hear his ideas about how this might work. We had an enthusiastic conversation, not least because his idea was so welcome. Soon thereafter, I took up the idea of converting the $14,000 loan balance into a grant for the board's approval. The board readily agreed, and that's how a small grant to open the CBF Virginia office set off one of the great environmental advocacy success stories in Virginia.

The Richmond office opened on October 1, 1980, staffed by a young attorney named Jeter M. (Bud) Watson. Bud's enthusiasm for his new job made him many friends among other Virginia environmental advocates. His initial efforts included monitoring administrative and legislative activities related to the Chesapeake Bay. The CBF office became the first group devoted to improving and protecting the Chesapeake Bay in Virginia that was staffed, rather than volunteer-based. CBF's motto, "Save the Bay," soon became ubiquitous; its bumper stickers lined the fenders of automobiles throughout the eastern part of Virginia. Years later, recognizing how valuable the state office was in Virginia, the Bay Foundation opened state offices in Maryland and Pennsylvania.

VEE helped the Bay Foundation launch its environmental education programs in Virginia in 1980 with a $2,000 mini-grant. It supported a teacher workshop, using its canoe fleet to support ten field trips for students and teachers from rural school divisions such as Essex, Charles City, and New Kent counties and the town of Colonial Beach. Later, VEE helped CBF to connect with the Richmond-area Mathematics and Science Center and the Science Supervisor at the Virginia State Department of Education. Beginning in the Hampton Roads area, the Foundation staff took groups of elementary students out into the Bay on its small boat. The demand for these trips grew rapidly, and before long they became so popular that the state government began supporting CBF's education program financially.

At the same time, CBF continued expanding its Virginia Office, now including a director, a chief counsel, environmental educators, and a grassroots organizer. Joe Maroon, who would later succeed me as the Endowment's executive director, became the first full-time director of CBF's Virginia office. VEE continued its strong financial support for CBF in Virginia, and its challenge grants have helped them raise significant funds to support their Virginia operations, to the great benefit of all Virginians who support a clean Chesapeake Bay.

Key to this successful expansion and operation was Will Baker, who was named president of CBF in 1982. Baker's vision of how CBF could "Save the Bay" was far-reaching. By skillful relationship-building and fundraising, utilizing advocacy and persuasion, he steadily built the organization into the powerhouse it has since become. Its biannual report on the state of the Bay is considered required reading for anyone who cares about the Bay.

Some of Virginia's most prominent environmental leaders have led the Virginia CBF office. Joe Maroon, Ann Jennings, and Roy Hoagland all played significant roles in building CBF's effectiveness over the past quarter-century. Peggy Sanner does the same now.

At almost every important CBF initiative in Virginia, VEE was there with first-dollar funding and encouragement. It has been a rewarding investment that has paid many dividends over the years.[1]

When VEE began its work in 1977, the Chesapeake Bay's restoration was not a priority for state government. That started to change during Governor Charles Robb's term, when he established the Governor's Commission on Virginia's Future in 1982. This blue-ribbon commission had natural resources as one of its six focus areas. Former state senator FitzGerald Bemiss headed the natural resources committee of the commission and made the Chesapeake Bay a priority.

Senator Bemiss invited me and Senator Joe Gartlan's legislative aide, Susan G. Dull, to help him analyze what the current situation of the Bay was in Virginia and what might be done to improve its status. Working again with Gerry Bemiss was a joy. He was such a brilliant man: a successful politician, businessman, and student of history. His ability to think and speak clearly and succinctly was without peer among my acquaintances. "What is Virginia actually doing about protecting the Bay?" he asked in a variety of ways. "Who's in charge, who could be useful?" He felt strongly that he had an opportunity with the Robb commission to move Virginia toward substantive action to protect the Bay and its natural resources, and he was not shy about tackling it head-on.

He was not interested in vague ideas or opinions; he wanted us to give him facts and information, so that he could pull them together into something worth recommending to the full commission. With our help, he wrote a compelling natural resources chapter for the commission's final report, laying out what Virginia needed to do to protect the Chesapeake Bay.[2] After quoting Article XI as a clear statement of Virginia's policy toward the environment, the chapter on natural resources notes, "We do not require a new or revised statement of public policy. . . . What we do require is a serious commitment to satisfy the intentions of our Constitution, backed up with intelligently designed programs and appropriate levels of funding."

Bemiss, in that one paragraph published in December 1984, sums up the decades-long goals of Virginia's attempts to implement its constitutional mandate. In his polite way, he goes on to suggest that Virginia's "conservation programs are neither appropriately designed nor adequately funded."

He concludes, "Virginia has reached the stage requiring a new level of commitment, new structures, and a clearer and firmer expression of our collective will." The report specifically recommended the establishment of a Secretary of Natural Resources. Two years later, Governor Gerald Baliles made good on that idea by persuading the General Assembly to establish this secretariat.

With respect to the Chesapeake Bay, the report outlined several problem areas: "We appear to be in the process of incapacitating the Bay for some time. One indication is the sharp decline in rockfish, shad, and oysters, which depend on clean water for their survival." The report takes note of the recent EPA study describing the present state of the Bay, which concluded "that our accumulating abuse of the natural system is affecting its productivity to a critical degree. . . . Nutrient concentrations have increased, mainly nitrogen and phosphorous. . . . Oxygen content of water in some parts of the Bay is being reduced and, in some places, eliminated. . . . Toxic substances are at high levels in localized bottom sediments, entering the Bay from point source industrial discharges and from nonpoint urban and rural runoff, particularly Baltimore Harbor and the Elizabeth River."

Senator Bemiss spoke from experience in commenting about the General Assembly's new ten-year commitment to fund a Virginia Chesapeake Bay Plan: "The record shows how easy it is for a legislative body to make grand commitments in response to a surge of enthusiasm for a popular cause of one moment only to abandon them at the next moment to favor something new and apparently more popular." I had often heard him express this reality in simpler terms: "I shall keep repairing my old Duesenberg while you play with your new Bentley."

This prescient report recommended that "particular attention should be given to the condition of individual river basins that contribute damaging quantities of nutrients and other chemicals from surface runoff and from inadequate industrial and municipal treatment facilities." The report never mentioned watershed improvement plans (WIPs), but this approach became the major new strategy adopted by the Bay Program some years later.

The main takeaway from the report was that restoration of the Chesapeake Bay must involve all the commission's natural resources recommendations on water quality, land use, and waste disposal. These three are naturally and inevitably linked, and that linkage "ties the Chesapeake Bay's conservation and use to our onshore use of land and water."

Senator Bemiss was a longtime champion of better land use planning and regulation in Virginia, and early on he posited the connection between onshore land use and water quality. It took until 1988, with the passage of the Chesapeake Bay Preservation Act, to fulfill that goal in law.

The report's other Bay recommendations called for Virginia to develop its own Bay restoration strategy in concert with the other states involved; to set up such a program regardless of the availability of federal funding; and again presciently—years before the Chesapeake Bay Agreement of 2000 called for it—"the Commonwealth should accelerate the development of its scientific data gathering and interpreting procedures, which are essential to an effective fisheries management program." Soon thereafter, the Endowment's board adopted the commission's recommendations for the Bay as its grant-making agenda for the next several years, first to publicize the report's recommendations and then to help fund the implementation of them. In the years immediately following the release of the report on the future of Virginia, the Endowment emphasized making grants that would support the commission's recommendations for improving the Bay. VEE continues to have a focus on the Bay even today.

Meanwhile, forces at the federal and state levels were gathering to improve the quality of the Bay. In 1983 a federal-state partnership to save the Bay was formed. The federal Environmental Protection Agency and the states of Maryland, Pennsylvania, Virginia, and the District of Columbia created the Chesapeake Bay Program. Initially, the administrator of the EPA was designated the chair, along with the governors of Virginia, Maryland, and Pennsylvania and the mayor of Washington, DC, comprising the Chesapeake Executive Council. Also in 1983, they signed the first-ever agreement to protect the Chesapeake Bay. Only a single page, it was more aspirational than operational. It had no plan, no strategy, no consideration of the role of science and technology in reducing pollution or managing the fishery. But it was a start.

The role of science in the formulation of Chesapeake Bay strategies would become significant in the years ahead. Politicians and policy makers long for precision; they want to know if an investment of public funds will produce a desired result. Scientists, however, offer conclusions from their studies with varying degrees of confidence. For example, a scientist might say that a certain result is likely with a 95 percent degree of confidence.

This gap between science and policy is part of what makes public science policy so interesting for those of us who have ever studied and practiced it. When I was in the Air Force, I studied for a master's degree in public science policy administration at the University of New Mexico. UNM was one of the first universities in the nation to offer a program in Science Policy Administration, and while it was based partly on the country's experience with NASA putting men on the moon, the lessons and practices are just as valuable to fisheries management policies.[3]

This information gap began to close when Jerry Baliles was elected Governor of Virginia in 1985. Governor Baliles had long experience with environmental issues. Among his priorities was improving the Chesapeake Bay, and he quickly set out to act.

First, he persuaded the General Assembly to establish the cabinet position of Secretary of Natural Resources, putting environmental matters on an equal footing in the government with other major areas. He selected an experienced environmental attorney, John Daniel, to fill the post. The next thing he did was call for a revision of the 1983 Bay Agreement, which he found to be full of pieties for saving the Bay but lacking any plan, priorities, or promises to pay for them. Also, the EPA administrator was designated the permanent chair of the Chesapeake Executive Council, which consisted of the signers of the Agreement. Governor Baliles suggested that the chair be rotated annually among the signatories, and he volunteered to become the first state-level chairman of the group.

Governor Baliles persuaded his Executive Council colleagues to develop a new, comprehensive Bay Agreement. He also had the good sense to revitalize the Chesapeake Bay Program's Citizen Advisory Council (CAC) and actively solicit its advice and input for developing a new Bay Agreement. He appointed experienced and knowledgeable Virginia citizens to the Chesapeake Bay Citizens Advisory Council to advise the government agencies on the work. This was a major salute to the many people outside the government who had an interest in protecting the Bay, and I was among those appointed. After a year of hearings and meetings, the CAC persuaded the Executive Council members and their staff to adopt nearly all its recommendations for managing the Bay into the new Agreement, to be called the 1987 Bay Agreement.

The CAC chair during this development phase was Edwina Coder of Pennsylvania, who was the perfect person needed to assert the citizens' demands for a Bay Agreement with "teeth." Coder was a strong and diplomatic leader who rallied the CAC's constituencies by holding public meetings to solicit advice from all those who shared the goal of a clean Chesapeake Bay. These hearings, managed by the nonprofit Alliance for the Chesapeake Bay, led by the extraordinary Frances Flanigan, provided the CAC with fresh ideas about eliminating toxic substances discharges into the Bay; managing the fisheries that lived within it; and dealing constructively with the growing number of people moving into the Bay region and their attendant need for housing, roads, and other necessities that accompany such development. The CAC also made it plain that it wanted the new Agreement to assure an ongoing role for citizens in its implementation and that additional public access points to rivers and the Bay needed to be provided. All these ideas were formally presented to the Chesapeake Executive Council. Edwina Coder later estimated that about 90% of CAC's recommendations were incorporated into the new Agreement. Much of the credit for achieving the 1987 Agreement is due to Edwina Coder and Fran Flanigan.

The 1987 Agreement was a major advance over the 1983 document. Its numerous recommendations for action were specific and ambitious. For the first time it included an entire section on managing the living resources of the Bay (its fish and shellfish). It also called for, per the CAC recommendation, the "zero discharge" of toxic substances into the Bay from all sources, a goal, decades later, that is still more promise than performance.

Money to accomplish the goals of the 1987 Agreement was scarce at both the state and federal levels. The EPA established a Bay Program Office in Annapolis, but its commitment of money fell far short of what was needed to implement the agenda in the Agreement. Progress was slow and after a few years it was obvious a new commitment and agreement were needed.

In the year 2000, a new Chesapeake 2000 Agreement made it a priority to manage fisheries' harvest levels "to maintain their health and stability and protect the ecosystem as a whole." To accomplish this goal, the Agreement called for a new multi-species, ecosystem-based management plan to be developed by 2005.

The Chesapeake 2000 Agreement's call for this new plan was historic and unprecedented. Up to that point, fishing had been done in a largely data-free environment. The Agreement's approach called for a sustainable, evidence-driven management approach. This subject is treated in detail in chapter 20.

As the 1987 Agreement made clear, establishing a constructive connection between what happens on land and its effects on water quality was necessary. Virginia had taken some small steps in that direction with the 1972 Tidal Wetlands law. The Virginia Erosion and Sediment Control Act of 1973 set the table for what is now the largest Clean Water Act problem, controlling "nonpoint" sources of pollution from farm, field, and forest runoff after rains. But it would not be until 1988, after an eighteen-month round of negotiations facilitated by a new mediation institute that the Virginia Environmental Endowment had established, before the General Assembly enacted the Chesapeake Bay Preservation Act. With the passage of that law, the legal and regulatory connection between land use and water quality was finally established.

The Chesapeake Bay Program today is an enormously complex undertaking involving hundreds of moving parts and projects and many millions of dollars annually. The commitment to cooperation among the federal government and the several states that affect the quality of the Bay has been documented in additional agreements over time and is now being driven by a 2025 deadline to achieve the goals of Bay restoration.[4] Even so, the cleanup will have to continue for many years.

# 13

# Clean Water and the Growth of Environmental Advocacy

TODAY WE are fortunate to have hundreds of environmental groups throughout Virginia and the Chesapeake Bay region, many of which VEE helped to start by providing funding to hire professional staff and support volunteer efforts. We also provided funds to many local river protection groups and local land trusts. Virginia's current network of environmental advocates is extensive, experienced, and effective. The results these groups have achieved in recent decades are numerous and lasting. That may be the Endowment's most important legacy.

It is hard to imagine now, but when VEE started in 1977 there were hardly any environmental advocacy groups in Virginia. The few that existed were made up of volunteers who were earnest but who had little success in persuading governors or legislators to their point of view. There was no Chesapeake Bay Foundation Virginia office, no Southern Environmental Law Center, no University of Virginia School of Law Environmental and Regulatory Clinic, no one working in Virginia from the Environmental Defense Fund or the Natural Resources Defense Council. There was no environmental group that had the capacity to participate in proceedings before a regulatory agency.

During the fall of 1977, I had a conversation with Louise Burke, a lovely woman who was the spokesperson for the Conservation Council of Virginia, a loose coalition of dozens of environmental and conservation organizations in Virginia composed of volunteers like Mrs. Burke. If you can imagine your favorite older aunt, that was Louise—as kind and sincere and genuine as anyone I have ever known. I had known her for years and admired her dedication, optimism, and persistence. She believed strongly in the rightness of protecting and improving the environment.

We were in my tiny office in downtown Richmond, and she sat across from me and leaned in, her coke-bottle eyeglasses reflecting the glare of the fluorescent lights in the ceiling, a tentative smile emerging, resting her hands on the old-fashioned purse resting on her lap, and said, "Jerry, we need your help. VEE can help us make a difference at the General Assembly."

"What do you have in mind?" I asked.

"The Conservation Council wants to hire an executive director. This is too much work for a volunteer to do well, and we think we'd all be better off with a paid staff to do the lobbying, put out regular updates, let the legislators know we want a clean environment, and rally the troops."

Communicating what was happening during the General Assembly session was more difficult back then. There was no Internet and no email, and there were no cell phones. I could appreciate her problem, because I had spent the previous seven legislative sessions lobbying on behalf of my previous boss, the governor of Virginia. Legislative monitoring was 24/7 and required a lot of legwork to keep track of conversations and shifting plans, and to read every bill for hidden agendas, sudden turns of fortune, and changing alliances.

Taking up the Conservation Council's request at our next meeting, the discussion turned briefly to the Endowment's own role in lobbying the legislature. The board chose early on not to get involved in directly lobbying the General Assembly, or the Congress for that matter. This decision rested on a couple of principles that have long guided the Endowment's activities.

The first was articulated by Judge Merhige during the proceeding when he caused the creation of VEE. He made it clear that the Endowment's board members were to be independent, both from one another and from other organizations: "I know individually they are all independent . . . collectively they are going to be independent." In many of our subsequent conversations with the judge, he made it clear that board members and I should avoid situations that might impede our impartiality and independence, such as serving on the board of a group we might fund.

The second was best stated by Judge Henry MacKenzie Jr., one of the first board members named by Judge Merhige. The board decided that it would rather try to promote mediation than fund lawsuits, because, in Judge MacKenzie's words, "It's not our job to choose sides." Indeed, in VEE's first annual report, released during the initial months of organizing our operations, the board made its thoughts known about its role in advocacy:

Advocacy Is Not for Us:

Another mistaken idea the Endowment would like to clear up is this: unlike many groups that have the word "environment" in their names, the

Endowment is not an advocacy organization. It takes a moderate, constructive approach to problem-solving. Instead of being a lobbying group, it sees itself in the midst of environmental activities. It can help various branches of the government, industry with its many interest groups, and citizen groups with their variety of concerns. Frankly, the Endowment believes it is in a unique position to build bridges of communication among all the different groups interested in Virginia's environmental quality. The Endowment has the perfect opportunity to stimulate cooperation and communication. And when problems arise, it has the means to help solve them.

A few years later, when it became clear that good intentions weren't enough, the board came to view *funding* advocacy as a useful complement to scientific and policy research. The board thought that the fledgling environmental movement could, if it organized its members well, advocate for environmental laws and policies more effectively than VEE ever could. We decided deliberately that the Endowment would refrain from lobbying or endorsing lobbying efforts and focus instead on supporting groups we helped to start, such as the James River Association, Chesapeake Bay Foundation's Virginia office, the Virginia Conservation Network, and VIRGINIAforever. Time has since demonstrated the wisdom of this approach.

This approach, of funding advocacy as well as education and research, has worked well. Indeed, the help we provided to so many new environmental advocates over the decades, and the successes they have achieved, constitute for me one of the Endowment's greatest legacies: the development of a permanent and powerful environmental movement in Virginia.

## James River Association

The James River Association, formerly the Lower James River Association, is a good example of how this movement developed over time. For VEE, our relationship began with lunch.

A member of the all-volunteer board of directors of the Lower James River Association (LJRA) called me up and invited me to lunch at the finest restaurant in Richmond, La Petite France. Association board member Paul Murphy was a vice president at Reynolds Metals Company and close advisor to the company CEO. Like many of his LJRA board colleagues, he lived along the James southeast of Richmond, and he invited another director, John Curtis, to join the lunch.

La Petite France was owned and operated by master chef, Alsace native, and French Legion of Honor recipient Paul Elbling and his wife, Marie Antoinette, who also were generous philanthropists. It represented the height of classic French cooking and style with its impeccable service and elegant decor and flowers. When you entered this gracious space, from Marie's warm welcome to the tables meticulously set with fine china on white linen and silver cutlery and the surrounding walls covered in a handsome forest green, you were immediately transported to a different world.

Wisely, I let Paul Murphy make some suggestions for this sophisticated meal. Chef Paul made a lobster bisque that I still consider the best I have ever enjoyed. The balance of flavors from the lobster, the cream, the stock, and no doubt some French culinary magic makes it memorable today, even though Paul and Marie retired and closed the restaurant years ago after a thirty-six-year run. Lunch was delicious, and I also enjoyed learning about why the James River Association cared about the river and the threats they saw confronting it.

Whatever the expense to my hosts—we were there for a couple of hours dining leisurely and well—the lunch "cost" the Endowment $10,000 ($30,000 today), becoming the first of many grants to the James River Association over the decades since then. VEE gave the Lower James River Association the money to hire its first professional staff member, Patti Jackson, who did a great job of making the group an effective advocate not just for the James but also for the state's environmental laws and rules more broadly. In the world of philanthropy, this kind of a grant is called "capacity building." In plain English, we believed it meant simply giving the grantee organization the ability to do its job effectively. Subsequently, we helped the James River Association develop river conservation plans and educational programs and start the James Riverkeeper program, among other initiatives.

Today the James River Association remains a great success story, thanks to the leadership of its current president, Bill Street. It has an annual budget of almost $4 million and a staff of twenty-seven covering several advocacy, conservation, and educational programs that span the entire James River from its origins in the mountains to the mouth of the Chesapeake Bay. Today the James ranks as the cleanest major tributary of the Bay. The James River Association has become the major voice for protecting the river and conducts programs for community conservation, education, shoreline buffers, and watershed restoration. One symbol of how far the James River improvement has come is the return of the Atlantic sturgeon, an ancient fish whose appearance has changed little since dinosaurs roamed the earth. They can be more than twelve feet long and weigh as much as eight hundred pounds. Though they are still listed as an endangered species, sightings as they make their annual spawning migration up the James River in late summer have increased.[1]

## The State of the Rivers Report

By the late 1980s, other groups were forming in the Rappahannock, Shenandoah, Dan, Clinch, and Elizabeth River regions. A serious movement to protect local water quality was building across Virginia from its northwest to its southeast and from the eastern seaboard to the southwest mountains. Each of these groups was started by volunteers, concerned citizens who were upset about water quality conditions in their neighborhood river. Few of them were knowledgeable about the technical aspects of clean—or dirty—water when they started investigating local conditions and asking questions of state water regulators.

State and federal agencies have the prime responsibility for monitoring water quality conditions—and enforcing them. The Clean Water Act requires that states monitor and document the condition of rivers and develop lists of various categories of water, from clean to "impaired." In Virginia at that time, the State Water Control Board staff only inspected any given segment of river once every three years, and only performed limited sampling when doing so. Things have improved a bit; now the sampling is done every two years.

Comprehensive, easily digestible information was rare until Bill Tanger, a businessman in Roanoke and an avid river enthusiast, decided to do something about it. He visited me in 1996 and, showing me maps of river basins in Virginia, demonstrated that it was nearly impossible for the public to know the water quality condition of these rivers. He planned to develop a report that would be accessible and understandable to the public, containing both narrative and pictorial information about the state of river monitoring in Virginia. His hope was that by doing so he would raise awareness and encourage more citizens to become active in protecting and restoring rivers throughout the Commonwealth. We decided to help him.

A clearer picture of the quality of Virginia rivers emerged in January 2001 with the publication of the first (and so far, only) *State of Our Rivers* report, published by Bill Tanger and the Friends of the Rivers of Virginia (FORVA). FORVA is a statewide coalition of twenty-eight river conservation groups. The report received major funding from the Commonwealth of Virginia's Chesapeake Bay Restoration Fund, the Virginia Department of Environmental Quality, Philip Morris Companies, and the Virginia Environmental Endowment.

Virginia newspapers summed up the report with headlines such as "Report: Pollution Up in Virginia's Rivers," "Reports: Meant to Be Murky?," and "Report: Virginia's Water Isn't Getting Cleaner."[2] According to a follow-up article in the *Times-Dispatch* a week later, the Virginia Department of Environmental Quality's own report considered 3,770 miles of rivers to be impaired.[3]

The FORVA report was a comprehensive answer to several questions: What is the condition of Virginia's rivers? Is the water safe for drinking? Swimming? Boating? If it is not safe, what can we do to help fix it? Who can we contact? How do we make a difference? Considering that Virginia has over 49,000 miles of rivers and streams in twelve major river basins, one can appreciate why sampling one site in some rivers every couple of years is insufficient for maintaining water quality.

The report also makes clear that while information about rivers can be found in federal and state sources, it is not presented in a "user friendly" form. As required pursuant to the Clean Water Act, every two years, the Virginia Department of Environmental Quality (DEQ) issues the 305(b)/303(d) *Water Quality Assessment Integrated Report,* known as the "Dirty Water List," which enumerates those waters that are impaired or threatened with impairment. It describes water quality conditions, both good and bad, for the entire state. The 303(d) report details the Total Maximum Daily Load (TMDL), the amount of pollution a body of water can accept while still meeting Virginia's water quality standards, addressing both point and nonpoint sources. The six designated uses in Virginia against which water bodies are assessed are: Aquatic Life Use, Recreation Use, Fish Consumption Use, Shellfishing Use, Public Water Supply Use, and Wildlife Use. The 2020 303(d) list of waters needing TMDL study is eighty-three pages long!

As discussed in chapter 8, Virginia started making TMDL plans in 1990 that were intended to reduce pollution runoff from nonpoint sources. A decade after the NRDC report, *Poison Runoff,* environmental groups sued the EPA, saying Virginia's cleanup plans were lagging. The court agreed and by a consent order set firm deadlines, stating that 648 TMDL plans had to be developed by 2011. These river segment TMDLs are largely required on tributaries of the Chesapeake Bay, part of the "tributaries strategy" adopted by the EPA and Virginia to control poison runoff. Currently, a TMDL for the entire Chesapeake Bay is under development.[4] The idea behind the TMDL approach presumes there is a certain level of pollution a stream can "accept" and still meet water quality standards. But without frequent monitoring of water quality conditions—sampling one-third of state waters every two years seems statistically unreliable—how does the state know whether the TMDLs are effective?

This all took place against the background of citizens taking up the responsibility for monitoring water quality because the state was certainly not doing a good job of it. Citizens took a variety of ways to accomplish this, ranging from the Isaak Walton League's Save Our Streams biological monitoring to river groups taking water quality samples and bringing them to a lab for chemical analysis of the water contents.

The Virginia Environmental Endowment began to fund these river groups in 1983 when it provided funds to the Lower James River Association. Soon after, grants were made to "friends of the river" groups in the Shenandoah, the Rappahannock, and the Elizabeth. In every case the groups were organized and energized by their volunteer board members and others who shared their interest and concerns.

## The Elizabeth River Project

In 1991 the Elizabeth River, which borders the cities of Norfolk, Portsmouth, Chesapeake, and Virginia Beach on the south side of Hampton Roads, was widely regarded by the Environmental Protection Agency as one of the most polluted rivers in the Chesapeake Bay watershed. It harbored a legacy of toxic pollution, the abuse from four decades as one of the leading harbors of the New World: cancer-causing tar pooled in the bottom of the river from defunct wood treatment plants, polluted runoff from aging port cities built before storm drains were recognized as a pollution problem, and an absence of basic environmental stewardship practices. Fish in the river were covered in lesions and were discovered to have cancer. Virginia Institute of Marine Science researchers had been studying the fish in the Elizabeth for years; pollution levels were lessening with new regulatory action, but most people considered the river dead.

Action was needed, but data was required to document the extent of the poisoning of the river. We had made a grant to VIMS in 1984 to do field and laboratory research on biochemical responses of marine organisms to toxic pollutants in the Elizabeth River, an opportune setting for such studies. The results of VIMS's work analyzing fish kills from exposure to chemical compounds and heavy metal contaminants helped to lay the scientific basis for building a community-wide consensus to restore the Elizabeth River.

Marjorie Mayfield Jackson, a local reporter, was one of four people who came together over a shared concern about the Elizabeth. Also in the group was a scientist at the local office of the Chesapeake Bay Foundation, an environmental volunteer in Virginia Beach, and the head of the local Society for the Prevention of Cruelty to Animals. Together, they developed a goal of restoring the Elizabeth River and went about it in a systematic, organized, and effective way.

Marjorie Jackson describes how she first got interested in cleaning up the Elizabeth:

> I was a reporter at the *Virginian-Pilot*. As a reporter, you weren't supposed to get involved in stuff—you could write about it, but you were supposed to be objective and not get in there and make things happen. And I wanted to make a difference. So, I quit my job at the *Virginian-Pilot* and looked for people who would be interested in helping to clean up the Elizabeth River. We didn't know each other really. We were people who independently had come upon the notion that it would be a good idea to help that river.
>
> One was Mike Kensler from the Chesapeake Bay Foundation, who had started an office down here. And obviously this was a major tributary of the Chesapeake Bay; they thought they ought to be concerned about it. Robert Dean, who had started Clean the Bay Day—I had seen him quoted in the paper saying the next thing he wanted to do was clean up the Elizabeth River. And I'm like, "me too." So, I called him up. And then Sharon Adams, who was head of the local SPCA. She knew Robert. We gathered in her house.
>
> The first decision that we made is that this river—this urban river that's home to the world's largest Navy base, world's largest coal export facility, twenty-something ship repair facilities—we are going to be ignored as powerless little mosquitoes if we blame and sue and hold protests and all that, which was the model at the time for environmental work; it was all pretty adversarial. We decided that instead, we're going to need to be about partnerships and collaboration and working together, to make real headway. And we have followed that model ever since. It has proved truly game-changing for the health of this urban river: bringing together all the powers that affect the river to work collaboratively for positive change, rather than fueling conflict among them—the latter immediately causes potential partners to throw up barriers to being part of the solution.

Sharon Adams, the SPCA director, was a friend of mine from my lobbying days at the state legislature. She called me and said, "Jerry, I've got a great project for you! I want to bring some friends up to Richmond to meet you and tell you all about it, so you can support it and we can get started." I said, "Fine, come on up and we'll talk."

Marjorie, one of Sharon's colleagues, would later put it this way: "We came to you because we wanted to start with a survey. We didn't know if people would want to get on board, collaboratively or not. I mean, the river was in trouble and so important to the economy here and to the nation's military powers. We just didn't know, really, if there would be support for grassroots, community-led effort. So, we thought, well, we'll get a little money and see if we can find out if people care. So, we called VEE."[5]

I met the four intrepid river restoration folks at my office at the James Center in Richmond. Aside from Sharon, the others were strangers to me, so Sharon made introductions. They proceeded to tell me how bad a condition the river was in and that it was about time that somebody did something about it.

Marjorie quietly laid out their goals for eliminating discharges, building support, and educating the public about the problems. Their confidence and determination were evident, but this was a major restoration project they were considering.

I asked, "How do you propose to accomplish all this?"

"Well, we can't do it by ourselves," Sharon said. "We need buy-in from just about everybody—the local governments, the businesses, the school systems—because we all contribute to the pollution in some form or another. And we need to raise awareness of how bad things really are and build support for the restoration program."

After exploring those realities for a while and listening to their thoughts about what needed to be done, I asked, "So, how much do you think this is going to cost?" They looked at me as earnestly as anyone ever has, and said, "We're looking for a grant of $1,375."

"$1,375?" I said.

"Yes, that should get us going just fine."

Stunned by the disparity between the amount requested and the scope of the work ahead of them, I said, "Okay, send me something in writing, and I'll see what I can do." I had authority from the board of VEE to approve small grants up to $5,000, so if I were convinced by their proposal, I could make the grant quickly. After they left, I shook my head and thought, "They are really going to make this happen."

A week later, their proposal arrived requesting $1,375. I looked at it and thought, "What could they possibly hope to accomplish with such a small amount?" They planned to begin with a survey to find out what people thought of the river, whether they felt it should be cleaner, and what approach should be taken if a cleanup plan could be prepared. $1,375 seemed like enough to make that happen.

On January 31, 1992, I wrote the check and wished them all the best. Years later, I asked Sharon Adams, "Why was it important for you that the Endowment give you this grant?"

She replied, "Because the Endowment's grant gave us the patina of seriousness, relevance. The money was great, and we certainly needed it, because we were funding it out of pocket, but it pales in comparison to being able to say, 'The Virginia Environmental Endowment is investing in this idea of saving the Elizabeth River.'"

The survey of more than sixty different people from all walks of life validated the initial thoughts of its organizers—people wanted the Elizabeth cleaner, even the business leaders. And they agreed on one thing: don't point fingers.

Marjorie later explained: "I went to do the survey, and the people who cared the most were the maritime businesspeople. They told us the Elizabeth is the 'Central Park' of Hampton Roads, our lifeblood, and they wanted it cleaned up—but don't point fingers. They said don't focus on the past, the damage done, this is not about the past."

Reassured they were on the right path, the organizers of the survey formed a nonprofit to manage the restoration effort; they called it the Elizabeth River Project (ERP). In the next few months, they recruited a board of directors. Marjorie Mayfield Jackson became the first, and so far the only, executive director of ERP.

Momentum gathered. Soon the group gained an EPA grant, combined with a larger VEE grant as matching funds, for a day-long symposium to which they would invite local government officials, state officials, educational groups, teachers, the heads of the industries along the riverbanks, and scientists from the Virginia Institute of Marine Science and Old Dominion University who had studied the river's fish and pollution levels. The group believed if they could just get everyone in the same room and have some speakers describe the problems and what might be done, a consensus would begin to form for the task of restoring the river.

The initial symposium was also the kickoff for a Watershed Action Team that would meet for four years to examine the science and hash out differences over approaches, in search of those acceptable to all interests. The Endowment's initial grant to VIMS in 1984 was for research that analyzed the extent of the poisoning of fish by toxic chemicals and compounds, and it was carried out by Dr. Robert Huggett. Marjorie Jackson makes the connection between the scientific evidence and the consensus that needed to be achieved to bring about action:

> Dr. Robert Huggett, a scientist at VIMS [Virginia Institute of Marine Science], was on the National Science Advisory Board and our initial ERP board. He suggested we use a planning model he had seen work for the state of Michigan to address our first huge challenge: How do you get diverse interests to talk to each other and agree on anything? You know, here we are at Milepost Zero on the Intracoastal Waterway, with the world's largest Navy base, all these shipping interests, Norfolk Southern, garden clubs, and all these residents who live on the river. How do you get everybody to agree?
>
> At that time, it wasn't that long after the Clean Water Act (amendments of 1987), people were angry—it was a really hard time to get a good dialogue going, because the regulations were just coming into play and people were having to change their act. Some people in industries were even stepping down, retiring, because they didn't want to get blamed for the legacy at some of these sites.
>
> Bob Huggett knew of a model in Michigan on comparative risk analysis, which is a boring way to describe getting everybody together for marrying science with public values.
>
> Thanks to Bob we were accepted for an EPA grant that funded our planning effort, guided by this little office at EPA called the comparative risk planning branch. He took me to Washington to see the director, and we got a little grant from them, and they helped us a lot. They had at the time bright young staff that led you through this planning process of getting diverse interests, scientists and the public and the industry representatives, all to agree on what the worst problems were environmentally and what the solutions should be. And we went through that process for four long years.

The first restoration plan for the Elizabeth was underway.

Marjorie describes the celebration: "When we unveiled the plan, people gave a long, standing ovation—when CBS News's Charles Kuralt unveiled the plan with a brilliant presentation at Nauticus in Norfolk." Getting Charles Kuralt to be the master of ceremonies and lead presenter of the new plan was a great adventure, according to Sharon:

So, when the time came for us to launch the plan to the team, we went to Scott Harper, the environmental reporter for the *Virginian-Pilot,* but I knew we had to have something exciting. See what he thought.

We decided to ask one of the great storytellers of modern television, Charles Kuralt, well known by that time for his "On the Road" series, to be our emcee. He was a North Carolina boy, and he had the right accent. And I said, but how are we going to get Charles? I went to people who might know Charles, because we never had a relationship with him at all. A writer for the *Pilot* whom I had known forever knew Charles's brother and agreed to touch base with him. "Tell him Charles will love this story," I said.

Brother gives me Charles's phone number in Manhattan. I had two minutes to make my elevator speech, so I did my best. He kept talking, and I kept telling him what we needed. And I said, we got no money, but I'll get you to Norfolk and I'll put you up. But I can't do anything beyond that. So he said yes, and he agreed to come to Norfolk!

He did a brilliant job. You were there, you remember. It was great. Slideshow too (produced by Bill Cofer, then head of the pilots that guide ships into the harbor, from his years of photographing the river, and his wife Susan Cofer, who chaired the Watershed Action Team). Afterwards, we all went to dinner at the Marriott where he was staying. What a gracious man he was. He was such great company, and he picked up the entire dinner tab for a dozen of us.

And we stayed in contact until he died.

Marjorie remembered the exciting events of the following months: "Our staff went from two to eleven within a few months as people learned about the plan to restore the Elizabeth. There were banner headlines in *The Virginian-Pilot.* I think there were eleven articles in one week. People just couldn't believe 'they're going to clean up the Elizabeth River.' Couldn't believe community leaders had agreed to do it. And people couldn't wait to be part of it."

The Elizabeth River Project was one of the first environmental groups to use a collaboration model. ERP built relationships among the businesses along the river, the four local governments bordering the river, and many different constituencies living near the Elizabeth. A key ingredient to getting buy-in from so many people was ERP's nonjudgmental, forward-looking approach. As Marjorie described: "That was key for us, and still is. From that foundation point, we've grown what we call the River Star

Businesses program, which is the oldest continuous voluntary recognition on the Bay for industries to do their part for the health of the Bay, above and beyond the law. And almost all the major facilities on the river participate. They document their results. We have a banquet for them."

ERP also started an education program for children of the school districts bordering the river. It, too, has become a big success story.

The education program is aimed primarily at school children, but adults love it and learn from it too. It features The Learning Barge, a solar-powered 120- by 32-foot facility constructed from an abandoned river barge that was retrofitted into an environmental laboratory, where students can spend a day studying water quality, wetlands, and river habitat and learning how to conduct scientific studies. It is the world's first floating wetlands classroom and was launched in 2009. Big thanks are due to Dominion Virginia Power, which contributed $250,000 to the barge's construction. As Marjorie described it:

> The Learning Barge was designed by the University of Virginia's School of Architecture. We needed a way to reach kids and get them on the water, but urban kids often have never been to the riverside, even though they might live just blocks from it. So, Phoebe Crisman from UVA was doing some land planning with us about another project (an industrial cleanup alongside a port facility at Money Point in Chesapeake). And she was like, you know, you really ought to have a public component of these projects to let them see what you're doing with these big cleanups.
>
> And she looked out at the river and there was a barge. And she said, well, maybe a classroom on a barge. And she began to sketch the idea with a live wetland on board. Her students got enamored of it and won a dozen awards for the design.
>
> So, we had it built here at one of the shipyards. And then UVA spent two summers bringing thirty students here. They spent the summer here and built the classroom element, which was on top of the steel deck barge. It symbolizes the working river reclaiming its heritage as a living river, because it has live wetlands and is powered by sun and wind, and in non-COVID times that barge stays booked, 6,000 kids a year, and they move through six hands-on learning stations.

The Dominion Energy Learning Barge, named to recognize its lead donor, was christened on September 14, 2009. More than 1,300 students from nineteen schools visited it in its first two months; it is fully booked throughout the academic year and most years has a waiting list of teachers and classrooms ready to pick up on any cancellation.

The driving force, lead actor, and "royal leader" of the Learning Barge is ERP Deputy Director of Education Robin Dunbar, who dresses in meticulously detailed seventeenth-century period dress that she makes herself, to portray Princess Elizabeth Stuart, daughter of King James I of England (VI of Scotland), for whom the river is named. ERP's Learning Barge is an enormous success story, and its class schedule is filled every year.

Robin's words best show how special the Learning Barge is:

> The barge continues to book a year in advance. We have had years of support from NOAA (National Oceanic and Atmospheric Administration) that has allowed us to offer all Norfolk Public School's 4th graders annually to participate in a Learning Barge field trip and we also pay for the transportation. In addition, they participate in Wetlands in the Classroom and are recognized as Resilient River Star Schools. (There are more than 200 schools to have been recognized to date.)
>
> One of my favorite things about the barge is how excited the youth are to come aboard and how proud they are to have this vessel in their own backyard to learn about their home river. I have heard students say, "This is my favorite field trip ever!" "All I want to do is touch the Elizabeth River water!" "This is the first time I have ever been on a boat!" "I can't wait to touch a crab or a periwinkle snail." "I love that each of us get our own supplies to use!" "I want to come here every day."
>
> And the students write the crew letters sharing what they learned, what was their favorite thing they did and how they are going to help restore the Elizabeth River. It's like Christmas every time we receive a large envelope from the schools full of student letters. . . . [S]ome even draw pictures that are priceless.
>
> I also love that I have been able to combine my love for science and art. I am an artist at heart and the barge has served as a wonderful creative outlet for me. I pour my heart into the barge, and it never fails to give back.

With a lot of help from private and public agencies, ERP became a multimillion-dollar operation with extensive local partnerships and additional financial support from the federal government. Over the decades since, river water quality has improved dramatically, oyster beds have been planted, toxic discharges have been substantially reduced, and a delightful and effective environmental education program was launched for the long run as well.

Not only has the Elizabeth River Project been good for the river's restoration, but it has also been good for the whole region. In fact, this effort was the vehicle for getting the four cities of Chesapeake, Norfolk, Portsmouth, and Virginia Beach to officially collaborate on a joint venture for the first time. All the mayors signed a proclamation at a celebratory event to mark the occasion of this unprecedented collaboration.

The Endowment's seed grant has grown a millionfold since that first meeting. Today, the Elizabeth River Project leads the lower Chesapeake Bay in collaborative grassroots restoration of Bay tributaries. Bringing people together works.

After leading ERP for thirty years, Marjorie Mayfield Jackson sums up her thoughts on their home web page: "From my perspective, there's been a dramatic change in how people perceive the river and how much they care about it. People are enthralled by it. They never tell me it's dead anymore. They just want to know what they can do to help."

The biggest changes, in her view? "I think moving from raising awareness of the Elizabeth as a polluted mess that has to be restored, to acceptance that it is a realistic goal to say it can be restored. I think most people thirty years or more ago had written it off as a disaster, and I don't think anyone thinks about it that way anymore. I'd like to think that all kinds of people now think the Elizabeth is someplace getting better and it's something they can get out and enjoy and appreciate and keep and treasure, as opposed to discharging junk into it all the time . . . to me, that's certainly a big turnaround in everybody's attitude toward the river—we're empowered . . . and we can do it."

In terms of advice to others, Marjorie's view is straightforward:

I think—unlike when we first came to you with that small ask—I think it's the big ideas that you can get funded, that will inspire people, and that people will get behind.

Somebody said: make no small plans, they have no power to inspire man's soul. I think that is true. Our most inspiring current project is, we're planning to build a "Resilience Lab" on Colley Avenue in Norfolk, and the design has so inspired people that we have million-dollar donors already stepping up.

Now, if we had said, we are just going to need a little office space upstairs, downtown, how much money do you think they would give us then? People are hungry for the big powerful idea.

It may have been a small ask, but it was a big idea. The restoration continues, powered by collaborative participation and demonstrating how much can be accomplished by just "a few caring people," as Margaret Mead famously said.[6]

## Lynnhaven River NOW

The Elizabeth's success has provided a wonderful example to other people in the region. For instance, around 2005 or 2006 Harry Lester and Andy Fine came to see me. They were developers in the Virginia Beach area who were also committed to conservation, especially of the diminished oyster beds in the Lynnhaven River. I knew Harry from our service together on the Commonwealth Transportation Board. Harry knew Sharon Adams, who suggested he contact me. He called up and said he'd like to come to Richmond with his friend Andy Fine, because they had a new river group they were starting and wanted to pick my brain about how to go about it. When we met, I'm positive I mentioned the Elizabeth River Project as an example in Hampton Roads. I also told them about the Friends of the Rappahannock, the Friends of the Shenandoah, the Friends of the North Fork of the Shenandoah, and other groups like them who started with some initial funding from VEE. Harry is a smart, charming man, and he smiled and said, "Hey, we're not here asking for money. We aren't that far along yet!"

They thanked me and went back to Virginia Beach, and the next thing I knew, a new group called Lynnhaven River 2007 had been formed. Then they asked for money, which we gave them—I think it was a mini-grant of $5,000—and they were off and running. Seed grants like that, right in the beginning, are incredibly valuable to a new group's confidence, credibility, and ability to raise additional funds.

They've done a great job continuing to raise funds and have had an extremely positive effect on oyster rebound in the Lynnhaven. I went down there to see the signs of their progress in growing more oysters—we call that a "site visit" in foundation jargon—and went out on the oyster boat with Cam Chalmers. Looking over the side of his boat, on a warm sunny day, I was amazed at the clarity of the water and how many oysters were visible on the bottom. They said they grow about three million oysters a year, which they sell to restaurants, so that was a tangible accomplishment that they'd already achieved. The oysters taste good too!

Lynnhaven River NOW, the organization's new name, exists today to restore native plants and lost habitats such as oyster reefs and salt marshes;

to reduce sources of contamination in local waterways; and to educate and engage the community in restoring and protecting their waterways. The organization employs a staff of more than fifteen people and maintains partnerships with dozens of local, state, and regional organizations in the Chesapeake Bay area.[7]

VEE has also invested in helping to start other local river groups such as the Nansemond River Preservation Alliance (NRPA), which, mainly through volunteers, provides shoreline maintenance, oyster restoration, water quality monitoring, and public access and advocacy.[8]

I first met Dr. Elizabeth Taraski when she was Vice President for Development at Virginia Commonwealth University in Richmond. After moving to the Suffolk area along the Nansemond River, she contacted me about a group she had recently founded in 2009, NRPA, and wrote a successful proposal for a mini-grant to get them started. They accomplished a remarkable amount in their first year alone, growing from less than a couple dozen founding members working with one school to hundreds of members working with eleven schools, with more than sixty local partners.

Another good example is the Dan River Basin Association (DRBA). DRBA provides numerous opportunities to get people out in nature to learn about their environment; recreational activities such as building trails and conducting a variety of local events that spotlight the river; and water quality monitoring and cleanup activities.[9]

One more example of citizens rising to meet the challenge of protecting their special place is the Friends of the Lower Appomattox River (FOLAR). FOLAR is a relative newcomer to river conservation efforts, having been started in 2000 to create a more detailed green- and blueway plan concentrated on developing a twenty-three-mile-long park and trail system that would become public access—by foot, bike, or boat. FOLAR hired its first staff member in 2014. The initial focus of the group's work was the completion of the Appomattox River Trail as a world-class regional trail system. In 2019 FOLAR received official resolutions of support for the Appomattox River Trail master plan from all six jurisdictions along the lower Appomattox River and significantly expanded its collaborative-partner support network. Several sections of the improved Appomattox River Trail were opened, including the Hopewell Riverwalk, which received the Governor's Environmental Excellence Award. In 2021, with a grant from VEE, FOLAR published the first land conservation plan for the lower Appomattox River corridor.

FOLAR, NRPA, and DRBA exemplify something we have seen many times: people who care about their environment—their "place"—coming together and acting to improve it and protect it from pollution, impairment, or destruction. Their distinguishing characteristic is leadership, consistent and persistent leadership. Persistence is needed because it simply isn't possible to deal with environmental problems on a short-term basis; we must commit to the long run, trying a variety of approaches and convincing other people and partners to join in the effort. I believe that even a little bit of progress and some small victories are better than nothing and that if one approach does not work, the best response is to try another, even if it turns out to also be in error, because based on what we learn we can stop, correct course, reorganize, and move forward.

Groups we've funded have failed in one way or another from time to time, but my observation is that, in the words of the old Timex watch commercial, they "take a licking but keep on ticking." As risky as some of our grants are, we're providing the capital in hopes they can do what they say they're going to do. And I think it's a two-way street. Part of what attracts our support is the demonstrable leadership skills of their founders and membership, but we also put our faith in them. I've gotten some remarkable notes from people telling me how much of an impact on their career an Endowment grant made years ago turned out to have; we supported them early on as a graduate student or activist, and it made all the difference in empowering them to devote their lives to environmental work. From Harry Lester and Andy Fine at Lynnhaven River, to the group of four that started the Elizabeth River Project, to Elizabeth Taraski who started the Nansemond River Preservation Alliance, grantees have all been initiators willing to commit to getting something done.

The role that river groups might play in restoring the Chesapeake Bay is critical, because controlling pollution in the tributary rivers helps keep pollution from infecting the Bay. This strategy has been evolving into a regional approach since the 1987 Bay Agreement first identified the major goals of the cleanup effort. Yes, the problems are global, but there are many solutions that are local—and if you don't get people who live in their own environment to take responsibility for it, you can only make very limited progress. When people grasp that they have the power to protect their neighborhood, their place, they organize and get going.

# 14

# Land Conservation and the Growth of Environmental Advocacy

"LAND USE is a fundamental determinant of water quality," the first annual report on the state of Virginia's environment declared in 1971.[1] Virginia is blessed with abundant natural beauty, but keeping it that way doesn't happen on its own.

In 1971, President Nixon's new Council on Environmental Quality (CEQ) commissioned a report about the use of land. This report—*The Quiet Revolution in Land Use Control,* written by Fred Bosselman and David Callies—examined the land use laws of several states to learn how some of the most complex land use issues were being addressed, including problems of reallocating responsibilities between state and local governments. The President's recently proposed National Land Use Policy Act was intended to provide federal assistance for states to develop programs dealing with land use issues of regional or state concern.

I drove up to Washington to meet with the CEQ staff most involved in land use matters to learn how the proposed bill might affect Virginia. We were concerned about how large projects such as the siting of major power plants might be handled, because their land use effects go beyond the jurisdiction where they are located.

When I brought my copy of the Bosselman and Callies report back to Virginia in 1972 and started asking people what they thought of the federal law to help states "reallocate land use responsibilities," I received a chilly reception. Even though Virginia is a Dillon Rule state, where localities only have the authority that the state gives them,[2] it is difficult for the Commonwealth to impose its will on its localities' jealously guarded land use prerogatives. The idea of regional or state-level decisions about the use of land was not realistic. But as we saw in the discussion of SELC's role in land use and transportation, we have made some headway, nonetheless.

Since the 1971 Governor's Council on the Environment report recognizing the connection between land use and water quality, there has been a steady evolution in land use policy and regulation in Virginia. In 1972 Virginia enacted its Tidal Wetlands Act, in what was probably the first time that the state interposed its own standards on local land use decision-making to protect a vital natural resource. The law established local wetlands boards to carry out state criteria when local permits were sought to alter or destroy wetlands in coastal localities.

By the 1980s, the Chesapeake Bay was beginning to be recognized as the national natural treasure that it is. A multistate and federal Chesapeake Bay Agreement was signed in 1983 that launched what has now become an extensive, and expensive, program to restore the environmental health of the Bay. Virginia, recognizing that it had to intensify and strengthen the legal connection between the natural relationship of land and water, negotiated and passed a landmark law in 1988 whose goal was, once and for all, to impose an affirmative responsibility on local governments to manage land uses in ways that protected water quality in the Bay region.

The Chesapeake Bay Preservation Act officially recognized the connection between the use of land and the protection of water quality, at least for the coastal zone of Virginia. The act came about as a direct result of the leadership of State Senator Joe Gartlan and Delegate W. Tayloe Murphy Jr. and deliberations by the Chesapeake Bay Roundtable. This group of about twenty people was led by Jim Wheat and facilitated by the Institute for Environmental Negotiation. The makeup of the roundtable was representative of the various groups who had an interest in the regulation of the coastal zone and of land use. For eighteen months, environmentalists, local governments, home builders, developers, representatives of fishing and agriculture such as the Virginia Farm Bureau, and others negotiated until reaching consensus on a draft bill.

Delegate Murphy had been a member of the House of Delegates since 1982 and quickly became a champion of environmental legislation. In addition to the Chesapeake Bay Preservation Act, another major law he sponsored was the Water Quality Improvement Act, which annually requires 10 percent of any state surplus be used to benefit water quality. In 2002 Governor Mark Warner appointed him Secretary of Natural Resources, a position he was perfect for. One of the biggest accomplishments he oversaw during that time was the initiation of a major increase in land conservation, reaching their four-year objective of protecting 400,000 acres throughout Virginia. He was also a champion of increasing state spending on the environment.

The consensus that the Chesapeake Bay Roundtable achieved did not hold completely during the bill's journey through the Virginia General

Assembly. But much of what the roundtable hoped for was still in the bill when it passed, a remarkable achievement on the first try for such a controversial subject.

The Chesapeake Bay Preservation Act established a state agency to oversee the implementation of the program, which was to be carried out by a new set of local boards in each Bay area locality. The law extended and surpassed the previous authority in the Tidal Wetlands Act of 1972.

Localities beyond the coastal zone objected to being subject to such a law and its regulations, arguing that many of them were nowhere near the Bay. This resulted in a political compromise that Governor Baliles and the act's sponsors were willing to make to get the bill passed, limiting the act's jurisdiction to the eastern coastal zone—even though, for example, the Shenandoah River, 220 miles from Norfolk, flows into the Potomac River, which in turn flows into the Bay.

Local decisions have more to do with shaping the environment for years to come than most people realize. Local governments' land use decisions affect the quality of water, the number of wetlands, the status of naturally important areas, the provision of parks and recreation and nature centers, and the condition of natural resources. The Bay act established a limited form of state oversight of such decisions. Furthermore, local government officials take an oath of office that includes supporting the Constitution of Virginia. Thus, in assuming their office, they also are bound to support Article XI, "to protect Virginia's atmosphere, lands, and waters from pollution, impairment, or destruction."

In the years following the CBPA's adoption, the Endowment made grants to the Chesapeake Bay Foundation and to the Middle Peninsula Planning District Commission to help implement the Bay Act, which, as one might imagine, was seen as controversial by the localities who had to implement it.

## Land Conservation Easements

If you drive north from Richmond on I-95, when you reach the Fredericksburg area, you can turn northwest in the direction of Culpeper and Loudoun counties. Spread before you, you will see an abundance of scenic views of farms, fields, fences, and forests preserved now by landowners who used perpetual easements to limit their future development and maintain the rural character of the area.

For the most part, Virginia's land conservation strategy has been to encourage private protection. But it started out differently. Virginia was the first state in the country to establish a public agency whose purpose was to protect the outdoors. The Virginia Outdoors Foundation (VOF) was created in 1966 by legislation sponsored by State Senator FitzGerald Bemiss, with great help in drafting the statute from Richmond lawyer George Freeman, a partner in the firm of Hunton & Williams.

The Commonwealth of Virginia's land conservation easement program enables a landowner to establish perpetual protection of property and continuation of the uses and attributes that constitute the land's conservation value. In return for limiting those development rights, the landowner becomes eligible for financial benefits as well.

Virginia expanded its land conservation program by establishing the Land Preservation Tax Credit (LPTC) program.[3] Under this program, Virginia allows an income tax credit for 40 percent of the value of donated land or conservation easements. Taxpayers may use up to $20,000 per year through 2020 and $50,000 per year in subsequent tax years. Tax credits may be carried forward for up to thirteen years. Hundreds of thousands of acres have been permanently protected using these credits. This program is a classic example of how land conservation and self-interest can work together to benefit the landowner and the public. This model of using what amounts to a combination of public and private incentives to protect land has stood the test of time and is much more the "Virginia Way" than a top-down state plan dictating what is to be protected.

The Virginia Outdoors Foundation, the largest state land trust in the country, has since 1966 partnered with thousands of landowners to conserve and protect over 850,000 acres of farmland and forests in 109 counties and cities. Without the Virginia Outdoors Foundation, the only statewide public land trust in the country at its inception, I'm not sure we would ever have had much of a land trust movement. VOF had no partners at the local level until 1972, when the Piedmont Environmental Council was formed by Arthur (Nick) Arundel and B. Powell Harrison, residents of Loudoun County, at the time an exurb of the growing Washington metropolis. In the past fifty years, the Piedmont Environmental Council, a private land trust, has been involved in protecting more than 580,000 acres in its nine-county region from Albemarle County north to Loudoun County.

From our first grant to The Nature Conservancy to protect its world-class barrier islands off the Eastern Shore, VEE has been involved in land conservation. And because scientific information is necessary for discovering valuable natural lands to preserve, the Endowment in 1987 put up the first funds for The Nature Conservancy to start a Virginia Natural Areas

Registry. After TNC demonstrated the value of this special program, Michael Lipford led the effort to persuade the Commonwealth to bring the program into its own conservation efforts, as a means to use science to advance land conservation. The state named it the Natural Heritage Program, and it focuses on science-based conservation to protect Virginia's native plant and animal life and the ecosystems upon which it depends.

In 1987 VEE made a $30,000 grant to The Conservation Fund, a new organization created by Patrick F. Noonan in 1985. After leading the expansion of The Nature Conservancy since the early 1970s, Pat saw the need for a group whose mandate went beyond the science-based conservation work of TNC. Pat is almost always enthusiastic about his work, but I remember his particular enthusiasm about this new idea. "This revolving fund will not only conserve and recycle land and historic buildings," he told me. "It will also recycle and grow the money so we can use it over and over!" We thought that a new national conservation organization based in Virginia, as is the national Nature Conservancy, would make a fine addition to the Commonwealth's land conservation efforts.

The Conservation Fund is headquartered in Arlington, Virginia. TCF was designed to cast a wider net than other organizations, so as to include historic preservation as well as a broader set of criteria for land conservation. Our grant established the Commonwealth Revolving Fund for land conservation projects that balance environmental goals with economic needs. One of the first projects was the reclamation and restoration of a historic building in downtown Staunton, a town that was in an economic slowdown at the time. By purchasing, repurposing, and selling the rehabilitated property, Staunton got a boost and TNC got more capital to replenish the Revolving Fund. TCF repeated this process many times and grew that fund to be a major catalyst in Virginia for environmental improvement that also helps the local economy.

Since its inception, The Conservation Fund has acquired and protected more than eight million acres of open space, wildlife habitat, and historic sites in all fifty states, including over 78,000 acres in Virginia. Today its programs encompass land, water, wildlife, and climate, as well as finance. To quote Larry Selzer, TCF's president, about its unique strategy: "At the core of our conservation finance efforts is the underlying belief that it is possible to succeed in aligning social, environmental and economic returns—what we call 'the triple bottom line.' Our name—The Conservation Fund—is predicated on our experience that environmental protection and economic vitality are mutually inclusive."[4]

By 1990, land conservation efforts in Virginia were becoming more widespread in response to steady growth and suburban development in many parts of the state. Seeking to identify how we might be helpful to the land conservation movement, we commissioned a study by Dr. Jon Roush, whom I had known since the 1970s, to investigate what needed doing, where VEE might be helpful, and what course of action we should follow. Jon is an expert in land conservation who had, among his other roles, served as chair of The Nature Conservancy and written extensively about land conservation.

In his report he wrote that there was a real need for land conservation at the local level in the face of continuing land development. He recommended that we could have the biggest impact by focusing on local efforts to conserve and protect land. The bottom line was: "Help the handful of smaller groups, the ones just starting out."

His conclusions led us to invest in smaller local land trusts just getting organized. With some targeted grants, we could partner with the Virginia Outdoors Foundation and the national Land Trust Alliance (LTA) to train them in the best land conservation practices.

One of the first new land trusts we met was the Valley Conservation Council (VCC). Faye Cooper, a cofounder of VCC, came to Richmond in late 1989 to tell us about this new organization and its hopes and goals. Faye Cooper recalled that first meeting:

> I just will never forget that meeting, because I was so nervous. I had never made an approach to the head of a foundation. I mean, that was my first experience with that, but, you know, you put me immediately at ease, and it was a great conversation. I went away feeling confident; "confidence," I guess, is the word I'd use. And I was not confident going in. It just gave me the confidence to write a simple grant . . . I mean, I think it did a pretty good job of making the case that, hey, we need help here in the Shenandoah Valley. That was a big, big boost, not only to the organization, but to me professionally . . . and though I was involved with it, there were a lot of other people too.[5]

Faye was serving as the Director of Stewardship for the Virginia chapter of The Nature Conservancy at the time. She had heard about the Endowment because of our support of TNC. She grew up on a family farm near Staunton, a "Valley lifer." We had the Conservancy in common, and that was the icebreaker to our conversation. What Faye said about being nervous, I saw as passion and enthusiasm as well as commitment to conserving her sense of place, the natural landscape of the Shenandoah Valley. In 1989 this was an ambitious goal.

She was also chair of the steering committee of the Valley Conservation Council. Her meeting with the Endowment was her first attempt to seek funds from a foundation.

Faye Cooper's childhood in rural Augusta County was enriched by parents who gave her the freedom to run the fields and the forests and to fish and to hunt. The outdoors was her natural habitat. In her own words,

> Those kinds of childhood experiences really matter when it comes to conservation ethic-building. And that's another thing that concerns me now. I care deeply about climate change and addressing climate change through land conservation.
>
> The Richard Louv book, *Last Child in the Woods: Saving Our Children from Nature Deficit Disorder*—that book had an enormous impact on me. And it made me think about my own experience and upbringing. I know young people out there who do care. For example, you can go to any climate change rally, and there'll be young people in the crowd, but we need more. And my worry is that because of—whether it's parents being scared to death to let their kids out of their sight or kids being plugged in to their phones and the internet, they just aren't being exposed to nature the way we were growing up. We need to create opportunities for children to be exposed to the natural world and get them outside.

I asked Faye about whether a "sense of place" motivated her work and her decision to spend her life in the conservation field:

> "Sense of place," we use that term quite a lot to convey a message that here in the Valley we live in special places with a rural heritage that in many families goes back generations.
>
> A major event that affected my future involvement in land conservation was the construction of Interstate 81. It literally blasted through the middle of our farm in 1963, destroying the biologically diverse wetlands associated with Naked Creek, rerouting and channeling the creek off the farm and ripping through fields and forests I had grown to love. It was devastating to this ten-year-old, whose natural playground was destroyed.

The Endowment made the first of several grants to help the VCC get started in 1990. The grant enabled them to open an office in downtown Staunton and begin their outreach to landowners about the value of conserving their land for the long term. VCC's vision to conserve land, water habitat, and local culture also included promoting modern land use policies that value land conservation as well as land development, instead of solely for what local governments call "its highest and best use."

The Valley Conservation Council's strategy for cultivating easements on farmland involves reaching out to local farmers, holding meetings in community settings, and making friends up and down the Valley:

> We focused on farmers: working with farmer groups and federal soil and water conservation districts who of course work directly with farmers, trying to get them engaged. And we all always included at least one or more farmers on our board.
>
> That went a long way also to help make those connections. We've been strategic in getting more engagement with the farm community. And it is resulting in more easements on working family farms. We did a huge easement recently on a six-generation dairy in Augusta County. And we're being approached by more and more farmers.

Like many other organizations, they started with a small group of like-minded friends and neighbors, invited them to come together and share refreshments and stories, and then added more and more people to their number. They also were sure to inform their local and legislative representatives of their concerns.

I went out to the Valley to see VCC's program in action. I arrived in Staunton in time to have dinner at a local restaurant with Faye Cooper. Over a leisurely dinner we talked about how Staunton was reinventing itself: new restaurants, the new Shakespeare Theatre (a true replica of the original in Stratford-upon-Avon), historic renovations and reuse of older downtown buildings, and a new sensitivity to the Better Models for Development approach, a guide to more sustainable development that VCC had sponsored in connection with the Urban Land Institute in Washington and which VEE had funded. We talked about how we had become interested in protecting the environment, and even though our backgrounds were hardly similar—hers in rural Augusta County, mine in New York City—we had both observed destruction and pollution at an early age, and those events were driving forces in our adult lives and careers.

My own childhood experience was the pollution of the Harlem River by the discharge of raw sewage directly from an open pipe into the river just upstream and across from where I occasionally liked to swim. When I was ten years old, early on a bright sunny summer morning, I and my friend Jackie stood on a bluff overlooking a bend in the Harlem River at 225th Street and Broadway, not far from its confluence with the mighty Hudson River a few hundred yards to the west. While the neighborhood adults headed for the subway and buses to get to work, we dreamed of a quick swim near the shore. "What's that brown stuff going into the river," I said to Jackie, who was eleven. "That's raw sewage from all the houses around here," he replied, except "raw sewage" was not the term he used. We were annoyed that we couldn't go swimming that day. Years later, when I was discharged from the Air Force, I remembered the blatant pollution of my childhood, and determined that I would try to do something about it. Pursuing the goal of clean water has been my life's work ever since I moved to Virginia in 1970. I believed as a child instinctively that there is no such thing as a right to dump that stuff in the river, that it was wrong. Still do.

The next morning I accompanied Faye to visit a farm or two and get a sense of how VCC was reaching out to landowners about its ideas for maintaining a vital farm community while at the same time preserving the rural landscape, mitigating water pollution, and earning money via the land conservation easement program. Faye introduced me to Bobby Whitescarver, the local Natural Resources Conservation Service representative. He was the federal government's man on the ground, a farmer himself, and well known to local farmers. Bobby was an enthusiastic salesperson for the farming way of life, despite a variety of challenges ranging from weather to annual price fluctuations.

Faye made the point that "Bobby is our secret weapon. He was born here, raised here, lives here, talks like he's from here, has his own farm, and has instant credibility with people. And we finally figured out that he's our best entree into talking with farmers about what we're trying to accomplish." He described some federal and state programs designed to help farmers practice good conservation techniques and to supply funds for water quality protection measures. We toured a couple of farms whose owners had recently donated conservation easements. Standing there, it became apparent how the appearance of that land today would for the most part be the same to an observer fifty years in the future. And there was also the likelihood that the family who owned that farm would still own it. On such a beautiful day, walking the fields, driving the back roads, and hanging around with such dedicated people, it was easy to be optimistic about the number of acres that VCC might protect in the years to come.

Since 1991 the VCC has worked with communities and landowners across Virginia's greater Shenandoah Valley region, promoting land conservation and sensible models of growth. As a result, tens of thousands of acres of valuable farm- and forestland, streams and rivers, and historical landscapes are now permanently protected in VCC's eighteen-locality, 5,903-square-mile service area, which runs from Botetourt County in the south to Frederick County in the northern end of the Valley.

VCC was among the first land trusts in Virginia to seek accreditation by the Land Trust Alliance. This means that it follows the highest legal, environmental, and financial standards to hold conservation easements for landowners. As Faye Cooper put it: "I think we've grown and improved markedly because of getting accredited; that validation process at the LTA setup is rigorous. I mean, truly rigorous. It took us four years to get through that process. It was so worth it—so worth it, because it got the house in order, in terms of the whole nonprofit best practices and standards."

VCC's responsibilities do not end when an easement is recorded. Every year, VCC staff visits the landowners that it has assisted to keep the relationship strong and to monitor how things are going.

The growth of VCC's easement programs and its responsibilities for monitoring them require ever more innovations to keep up. Drones are now part of the process of monitoring; satellite imagery is too. Land conservation techniques are evolving to take advantage of twenty-first-century technology.

The Valley Conservation Council has evolved and leads the way for land conservation and permanent protection in the Valley. It also works with other groups such as the Alliance for the Shenandoah Valley. The Alliance's work complements VCC's by working with local communities to influence land use planning and transportation decisions. Together, and along with more than thirty other partners in the Valley and nearby, they collaborate in shared land conservation and water quality goals.

VCC currently holds fifty-six easements totaling 6,550 acres of land. In addition to that, VCC in 2022 transferred another 963.5 acres to the National Park Service. As of March 2022, the total of all land-trust easements in VCC's eleven-county service region equals 213,650 acres. VCC assisted in obtaining almost half of those, by way of its education and landowner outreach program.[6]

Working with the Virginia Outdoors Foundation and the Land Trust Alliance, VEE has helped many local land trusts in Virginia achieve formal accreditation of their programs. Land trust accreditation is a mark of distinction, showing that a land trust meets high standards for land conservation. It sends a message to landowners and supporters: "Invest in us. We are a strong, effective organization you can trust to conserve your land forever."

We also helped start VaULT, the Virginia United Land Trusts, the statewide association of about thirty local land trusts, and funded its first annual conference in 2007. It is fair to say that VEE provided crucial help to Virginia's land trust movement overall. This is highly leveraged grant-making, because a small grant to a land trust enables it to spend time talking with landowners and showing them the benefits of donating conservation easements on their property. This is a win-win, because the landowner gets to keep the property and get a substantial tax credit for the easement, while the public gets the benefit of having this land preserved in perpetuity.[7]

The Virginia Conservation and Recreation Foundation, where I once served as vice chair, was renamed in 1999 as the Virginia Land Conservation Foundation and is an important part of the Commonwealth's land conservation efforts.[8] Its duties include to "establish, administer, manage, including the creation of reserves, and make expenditures and allocations from a special, nonreverting fund in the state treasury to be known as the Virginia Land Conservation Fund,"[9] to help fund protection of the Old Dominion's natural landscapes. This fund annually distributes millions of dollars for land conservation purposes: "Grants are awarded to help fund the purchase of permanent conservation easements, open spaces and parklands, lands of historic or cultural significance, farmlands and forests, and natural areas. State agencies, local governments, public bodies and registered (tax-exempt) nonprofit groups are eligible to receive matching grants from the foundation."[10]

As part of a major push to conserve land in Virginia, during the period from 2002 until 2010 Governors Mark Warner and Tim Kaine permanently preserved more than 800,000 acres in conservation easements. This was a tremendous achievement that involved many state agencies and nonprofit partners, and many VEE grantees were a part of the effort.

Another land conservation organization with a more recent history that is making tangible progress in protecting Virginia's landscapes for the long term is the Capital Region Land Conservancy (CRLC).

## Capital Region Land Conservancy

The mission of the CRLC is to conserve and protect the natural and historic land and water resources of Virginia's Capital Region for the benefit of current and future generations. CRLC is the only land trust devoted specifically to conservation within the counties of Charles City, Chesterfield, Goochland, Hanover, Henrico, New Kent, and Powhatan as well as the Town of Ashland and City of Richmond. CRLC is proud to have conserved more than 13,000 acres of land in this special region.

The Capital Region Land Conservancy protects lands through voluntary conservation easements and by acquiring land through donation or purchase: "Our projects cover farmland, forests, historic battlefields, rivers and streams, and vulnerable ecosystems. We give great care to projects from small family farms to monumental undertakings like protecting Malvern Hill Farm."[11]

## Sustainable Development

One ambitious but unfocused and therefore unsuccessful effort was VEE's "sustainable development" grants program. The concept of sustainable development had been around since the early 1970s—and some would argue since Teddy Roosevelt was president, although he used different language to promote both conservation and wise use of natural resources. It is a hard concept to define consistently, and the results are difficult to measure. It was not our best work.

As of 1993, we had made several grants to the Environmental Law Institute (ELI) in Washington, DC, to perform a valuable analysis of public and private environmental and development laws, programs, and practices in Virginia, which was published as *Blueprint for Sustainable Development in Virginia*. We had also set up an advisory committee on this subject, comprising business, political, conservation, and academic leaders, who advised ELI during its research. *Blueprint* provided a foundation for a sustainable development strategy for the Commonwealth in pursuit of the mandate in Article XI. Unfortunately, it did not attract much interest.

During the period from 1990 until 1995, the Commonwealth also studied sustainable development in its Commission on Population Growth and Development, cochaired by Senator Joseph V. Gartlan Jr. and Delegate W. Tayloe Murphy Jr. The General Assembly established the Commission in 1989 after the successful passage of the Chesapeake Bay Preservation

Act and charged the Commission with "studying and developing a vision for the future of Virginia."

The Endowment, seeking to find out if some of the goals outlined in *Blueprint* and by the state Growth Commission could be quantified in some ways, made a series of grants aimed at developing indexes of environmental quality. One grant was to George Washington University, which had a satellite campus in Loudoun County. GWU and its team labored to construct measures of environmental quality for the Loudoun County planners and supervisors to consider. Another grant to a local business association in Roanoke attempted a similar effort, to define and measure environmental quality in the Star City. Neither process lasted beyond the grant periods. Local elected officials must be on board for such projects to have even a chance at success.

By contrast, the long-term commitment by the Valley Conservation Council to conserve farmland and promote better models for development offers a working example of how to go about the business of sustainable development.

Overall, despite the inherent difficulties in defining and achieving sustainable development, the Endowment's support of land conservation has in fact helped to preserve tens of thousands of acres of scenic agricultural and forestal land. It is tangible, measurable, and a result that potentially will last in perpetuity.

# 15

# The Virginia Conservation Network

THE NUMBER of environmental nonprofit groups increased greatly in the late 1980s. This was a useful development in the long-range effort to protect Virginia's environment and made apparent the need for an effective statewide network of environmental advocates to work together on a common agenda to protect the environment.

The Conservation Council of Virginia (CCVA) was a statewide network of environmental groups that served as a lobbying arm for approximately one hundred environmental groups. These groups ranged in size from large to small, with many in between, some with paid staff, but most were all volunteers, including their leadership. They came from across the state, and their interests spanned a variety of topics related to the environment and natural resources. CCVA also had a supporting group called the Conservation Council of Virginia Foundation, a public charity to receive tax-deductible gifts to support CCVA.

In part because of the wide diversity of interests represented in CCVA, when it came time to lobby the General Assembly of Virginia, the conservation community often delivered mixed messages, because frequently there was no consensus regarding what the priorities ought to be, nor was there consensus on ways to respond constructively to both good bills and bad bills introduced by legislators. Some groups would support a piece of legislation while others would oppose it. Positions might switch on another bill. It was confusing and challenging for sympathetic legislators to try to pass legislation without united support.

This came to a head following the 1989 session of the General Assembly. Three members of the legislature—Senator Joe Gartlan, Delegate Tayloe Murphy, and Delegate Watkins Abbitt—came to see me at the Endowment's office, pleading for our help to make the environmental community get its legislative act together. Senator Gartlan and Delegate Murphy had been environmental leaders in the legislature since 1972 and 1982 respectively, and we had worked together many times. The previous year, Delegate Abbitt had waded into a long-standing policy dispute over ensuring there would be enough water in streams to sustain marine life despite water withdrawals for municipal water supply and agricultural uses. Adequate supplies for drinking water and for agricultural uses were already required during droughts, but there was no protection for natural resources such as fish and other aquatic life in streams. In addition, Wat Abbitt and I had served together on the Uranium Advisory Council with Elizabeth Haskell, later Virginia's Secretary of Natural Resources in the Wilder administration, and we liked and respected each other. In 1988 Delegate Abbitt, with the assistance of the Institute for Environmental Negotiation, developed, sponsored, and passed a new instream flow bill that made protection of aquatic life an equal consideration in calculating water withdrawals in shortage situations. It was an important environmental victory, accomplished in Wat's quiet way.

With respect to the CCVA, the legislators' complaint was that the environmental lobbyists were acting like soldiers standing in a circle and firing their weapons, wounding each other and undercutting their effectiveness. As a result, these legislators were, at least, confused and, at worst, totally frustrated by this lack of coordinated effort. The legislators asked the Endowment if we might convene perhaps twenty-five of the key groups, the ones doing most of the lobbying. We agreed and sponsored the first two meetings in the spring of 1989.

We held the meetings in comfortable facilities with ample room and beautiful settings. The first retreat was an overnight meeting on the grounds of Stratford Hall, the former home of the Lee family, now a historic house museum in Westmoreland County, Virginia, overlooking the Potomac River. Later that spring, participants enjoyed the hospitality of the James River Corporation's corporate retreat lodge along the James River. As befits a Fortune 500 company property, the facility was beautifully decorated, professionally staffed, and generously welcoming. The food and drink were plentiful and delicious. We tried to make everyone feel special, comfortable, and valued equally, the better to focus on the task at hand. The goal was to forge an effective collaboration among dozens of groups around a common environmental protection agenda. The attractiveness of these settings and the provision of plenty of delicious food and drink in the morning, noon, and night made for a convivial atmosphere that even the most resistant-to-change attendees wound up enjoying. Once again, the Institute for Environmental Negotiation staff proved invaluable as they facilitated discussions among sometimes contentious personalities.

Even the weather was bright and sunny, which put everyone in a good mood. The three legislators participated actively in those two-day meetings, arguing for consensus on major priorities among the groups so that they could present a unified front on legislation. Their recent experience with the environmental community's efforts at lobbying was that a divided community was a defeated community.

What began to emerge from all the conversations, formal and informal, was a sense of purpose larger than each group's perspective—environmental protection and natural resources conservation for the Commonwealth—and a better understanding of how others' points of view could be equally valid and useful, that they all had something to learn from each other and that by working together on some issues they could be more effective. The opportunity to spend serious time with each other was a new experience and allowed many to get to know each other much better. The level of discussion and engagement was unprecedented in the movement represented by CCVA.

During the third session in the fall of 1989, it was agreed that a new organization, building upon the Conservation Council and its supporting foundation, would be created to unify the lobbying among the members. They chose as its initial name the Virginia Environmental Network.

Overcoming organizational self-interest to create an effective network took several years and hundreds of thousands of dollars in grants to VEN by VEE. It also required a great deal of patience. Strong personalities, early clashes, and bruised feelings made for a lot of tacking back and forth in order to go forward. During those early years, there were disputes involving the role of the board of directors. Of course, there was some concern about fundraising for VEN in a world of limited funds and a therefore competitive fundraising environment. Leaders such as Patti Jackson of James River Association, Kay Slaughter, who was an attorney with the Southern Environmental Law Center, and Stella Koch, who was with the Audubon Naturalist Society, spent countless weeks negotiating the finer points of how the new organization would operate.

That first board was composed mainly of people who were either executive directors or senior staff in their organizations. That was one of the challenges, because representatives came to board meetings with their organizational role, not their VEN role, in mind. It was almost impossible for them to be an objective supporter of VEN and to figure out how to raise funds for it while already busily raising funds for their own organizations.

Over time, they successfully adapted to a new way of collaborating, and though it wasn't always pretty, they learned to treat each other as allies deserving mutual respect and cooperation. Throughout these growing pains,

VEE was determined to encourage the groups to work together, focusing on the issue that mattered most and presenting a common agenda to the legislature. It took until late 1993 for the boards of both the Conservation Council of Virginia and the Virginia Environmental Network to agree to merge the two organizations into one statewide network. Trip Pollard, an attorney with VEN member Southern Environmental Law Center, was recruited to draft the bylaws and articles of incorporation of what became the Virginia Conservation Network.

We supported VCN generously for the next several years, eventually granting a total of $350,000 to help it develop into an effective lobby on behalf of the environment for the people of Virginia. A much stronger organization emerged from the process.

A turning point in VCN's governance occurred when they decided to add independent directors to their board. Up to that point, the board had been entirely composed of representatives from member organizations. New board members were recruited, people like Cabell Brand, a businessman and philanthropist who had decades of experience serving on nonprofit boards and who brought a broader point of view to VCN. The independent directors could help by contributing money and in fundraising as well.

VCN staff and member organizations' presence during the General Assembly meant that committees considering environmental bills were always attended and monitored. Even committees without primary responsibility for environmental legislation.

An example of why this is important occurred during a meeting of a House of Delegates committee on local government. This committee's responsibilities dealt mostly with local government powers, duties, and responsibilities. On the committee docket that day was a bill the VCN had not noticed. It was a charter bill for the City of Richmond, that is, a bill that amended and reinstated the city's law. Many such bills are routine and noncontroversial. However, exceptions regularly occur in the legislature. When this bill came up for consideration, the committee's chair, Delegate Richard Cranwell, asked the VCN if anyone wanted to speak to it, perhaps even against it. Looking at each other, the representatives shook their heads "No." Delegate Cranwell suggested that they might want to object to it, because it would have eliminated the requirements of the Chesapeake Bay Preservation Act within the City of Richmond, a law that they had fought hard to establish just a few years earlier and one that local governments were at best reluctantly adapting to. In horror, the environmental lobbyists read the bill. "Yes, we object," they said, and the offending language was removed from the bill. "Read every bill" is the first rule of lobbying.

Today, VCN is a powerhouse and a diverse, well-coordinated conservation movement. It works in partnership with over 150 organizations encompassing four main programs: healthy rivers, clean energy and climate, land conservation, and land use and transportation. VCN is Virginia's lead partner with the Choose Clean Water Coalition, serves as the state affiliate of the National Wildlife Federation, and is a member of the Virginia Environmental Justice Collaborative.

Mary Rafferty has been VCN's executive director for the past six years. Her knowledge, enthusiasm, and general satisfaction with VCN's success makes for a good story in its own right. As Mary Rafferty explains,

> Our goal is to protect the environment, but what is unique about VCN? What I think will continue to persist is that it's all about relationships, and those will change; they're different today than they were thirty years ago— and will be in thirty more years. That need to ensure that individuals trust one another and support their vision and can work well together will persist.
>
> The network today has 150 different groups. That's a lot of people, and that's inspiring to think that many people are engaged daily in doing this work. And whether it's a local group that's doing tree planting in Southside Virginia, or the Chesapeake Bay Foundation that has hundreds of staff, they are all part of the network now. But you've got 150 unique organizations that are tackling environmental problems, kind of in their own vision, in their own way. And they have to trust that they understand what their role is and will accomplish their work. And they have to trust that another organization that's taking the lead on a given area is going to accomplish their role and do it well. When they need to work together, they can pick up the phone and talk to one another. When they disagree, they can tell each other that they disagree and find a compromise. And, when it comes to some of the hard things, like having to tell a new governor that you don't support who he picked for natural resources secretary, that you trust each other all enough to be like, all right, are we in this together?[1]

That's a big difference from the attitude in 1989 when the legislators asked for help in marshaling the environmental community.

Mary Rafferty goes on to say that

people in other states tell me often that they just can't believe there isn't something like VCN in every state; like, that just doesn't exist.

One of the coolest things that's happened over the last handful of years . . . it's little things like making sure that we're passing better pedestrian safety laws, which sometimes can be seen as like, well, okay, fine, better pedestrian safety laws, but it makes the streets safer. More people are willing to walk. People get out of their cars. Ultimately that means that we don't have to build as many homes outside of the city, because we've got people walking inside of the city, which protects land, which reduces our carbon pollution. Little steps like that help too. And somebody takes the lead on that. And we all trust the issue leader to do the work well, and then we support it as a network.

For example, one group will say, what we really want to work on is bicycle and pedestrian safety. And so they'll take that, they'll write a policy paper on how important that is. They'll come up with specific policies that they're going to advocate. And then VCN gives them . . . guidance and resources of, like, now here's how you find a patron and here's who you should talk to next. And here's when the deadline is to get your bill in. And here is the committee it's likely going to go through after all the bills have dropped.

Then, it becomes part of our briefing book for legislators. So somebody who's just trying to get a small—seemingly small—thing done is better equipped to get that small thing done. And they can put our name on it as well to say, yep, VCN now is on board, it's not only me, small bicycling group or local wildlife organization or whatever. I've also got the backing of, you know, the statewide environmental organization.

I think what Virginia has going for it now is that we really do have the infrastructure for statewide policy advocacy, and if there are local groups popping up, legislators get to see a local problem too.

VCN helps the environmental conservation community in Virginia speak with one voice. When the General Assembly convenes, VCN advocates for the common agenda its members have developed, tracks legislation, and coordinates partner advocacy for improvements to state laws and regulations as well as in funding state environmental protection and natural resources conservation programs.

VCN and its members worked effectively with the 2022 General Assembly to approve a large conservation budget, including an allocation of more than $87 million to the Water Quality Improvement Fund.

By the end of 2022, Virginia had created a 100 percent clean energy standard, a wildlife corridor plan, a variety of land conservation programs and tools, an environmental justice council, and historic investments in clean water funding. While changes in government leadership can try to change this great progress, VCN and its partners are strong enough now to limit the damage of such shortsighted attempts to weaken the measures put in place in recent years.

VCN has helped to build today's environmental movement into the force it is today. And they are not going away.[2]

# 16

# Telling the Stories

"WHY HAVE I never heard of any of these wonderful people?" Patricia Kluge asked. It was the spring of 1994, and Mrs. Kluge, a new member of the VEE board, was attending her first meeting. Up to this point, she had been listening closely and speaking only to vote to approve grants. The seven-member Endowment board was meeting in a comfortable conference room at the Boar's Head Inn outside of Charlottesville.

The grant application review is the heart of every VEE board meeting. That is when the board decides which proposals it can fund. The conversation frequently includes thoughtful questions and lively discussion of the proposals and what kind of results we could expect to see if we made the grant. The blending of voices in lively discussion improved every idea. The board's deliberations always clarified the nature and expected outcome of each proposal, and at this meeting they had just approved several new grants for Virginia projects.

The board members, casually arrayed around the table, gave Mrs. Kluge their full attention. An intelligent and well-connected person, Mrs. Kluge is a tall, striking woman with a lively sense of humor. The expression on her face now expressed both pleasure at the grants and total unfamiliarity with the grantees. She wanted to know more: "No, really, I'm serious. I think a lot of people in influential circles have never heard of these people, and they need to. Who's telling them how to tell their story?"

Mrs. Kluge spoke up in her elegant British accent, genuinely befuddled: "I think I get around Virginia quite a bit. I know a lot of people, and yet I've never heard of any of these groups. I mean, it looks as though they are doing great work, and good for us for supporting them, but they need to be better known," she continued. "More people need to know about the work they are doing on behalf of the rest of us. What happens after we make a grant? Do we announce it? Do they? How does the publicity part work?"

I thought it was an excellent question, and one the rest of us had not discussed in a while. We agreed that she had a point: We all knew, for example, the James River Association, the Institute for Environmental

Negotiation, the Southern Environmental Law Center, and the Virginia Conservation Network; she, and probably many other opinion leaders around Virginia, did not.

She asked, "Who does their public relations?"

"Well, usually we send out a press release to the media after we have notified the groups about the grant," I replied.

"Is that it?" she asked.

"Pretty much," I said, suddenly feeling sheepish about it.

Patricia expressed the view that we could help these people tell their stories, just the kind of "outside the box" idea the board liked. "We need to get a first-rate publicist working for these groups," she continued, "and I know just the one. He represents me, among hundreds of others. He's the best."

The board members were nodding and smiling as she explained what she had in mind: "These are their stories, not ours. They must be able to tell the public what they are doing to protect the environment. We can help them do that."

"Who is it?" asked Al Smith, a retired Virginia legislator from Winchester who knew Patricia quite well.

"Howard Rubenstein. He is in New York but has clients everywhere," she said.

Alson H. Smith Jr. was a businessman who owned a Tastee-Freez franchise and represented the Winchester area of Virginia in the House of Delegates from 1974 until his retirement in 1994. During that time this "good old country boy" became a powerful and influential member who knew not just how to get things done but also how to prevent things he didn't like from getting done. He also told me with a big grin, "You won't find my fingerprints anywhere on some things, either."

In the spring of 1994, Judge Merhige called me. "I've got a new board member for you, Jer."

"Who is that?" I asked.

"Al Smith. You know him, don't you?"

"You bet I do! I've known him for years, and I like him."

"He'll be in touch. I'll tell him to call you so you can get him up to speed."

In keeping with his previous choices of board members, the judge had yet again appointed someone with a reputation for excellence and effectiveness in their chosen field and with a wide range of knowledge and experience. It was my job to help them apply all that to the Endowment's environmental mission.

During his years in the legislature, Delegate Smith had not shown great interest in environmental legislation. Most legislators pick their principal topics and develop expertise in one or two areas. They learn to rely on other legislators for information about other subjects. Al Smith looked out for the State Troopers, among others. Many years later when I was a member of the Commonwealth Transportation Board, a State Trooper and I got into a chat. All I said was, "Did you ever know Al Smith?" and the trooper went on and on about what a great Virginian Al was and how he was their biggest supporter in the legislature.

When Al and I met to discuss the Endowment's work, he seemed both honored and puzzled that Judge Merhige had chosen him for the board. But he was enthusiastic and ready to plunge ahead. What he lacked in environmental knowledge was made up for in his charming way with the other board members who quickly came to enjoy having him with us. He had such a lively, curious way about him, and he was a quick study.

A few years earlier we had given money to a river protection group up in the Valley, the Friends of the Shenandoah River. These volunteers wanted to protect and improve water quality in the river and regularly took water samples to monitor its condition. The state Department of Environmental Quality, created officially in 1993 through the efforts of Natural Resources Secretary Elizabeth Haskell, encouraged their interest but couldn't rely on their sampling for regulatory purposes. The samples did not meet the "quality assurance / quality control" laboratory standards that the US Environmental Protection Agency (EPA) required. Neither the EPA nor the Virginia DEQ, however, did much monitoring of their own, as described in chapter 13. It was an odd situation of infrequent monitoring of water quality conditions by the government and plenty of people willing to volunteer to monitor water quality in their own backyards—who were facing an obstacle created by the government that should have been helping them to supplement its meager efforts.

I told Al about the problem. When it came to helping constituents, there was no match for Al's abilities. Even though he had retired from the state legislature, he maintained an interest in helping his former constituents however and whenever he could. Now, as an Endowment board member, and using his many contacts, he had another way to help those citizens.

"What do they need?" he asked. I told him that the Friends of the Shenandoah River needed to take their samples to a certified water quality laboratory for analysis before the state or federal agencies would accept their test results. I also explained that there wasn't any such laboratory near them and that it can be expensive.

"Let me give Jimmy Davis a call," he said. Dr. James Davis was the long-time president of Shenandoah University and a former colleague of Al's in the legislature. Dr. Davis had expanded Shenandoah from its base as a highly regarded music conservatory into a powerhouse university that had become a major economic engine of the Winchester area.

Al had also helped Dr. Davis with plenty of private fundraising and a decent amount of state help as well. If Al asked Jimmy to do something, Jimmy would find a way.

Not long thereafter, Shenandoah sent us a proposal to build a small but first-class water quality lab so that the volunteers' water samples could be professionally evaluated and certified. The results of that joint effort among the university, the monitoring group, and VEE continue to work well today.

Al Smith's modest reaction to this wonderful accomplishment: "Jerry, you're going to make an environmentalist out of me yet!" "Big Al" became a genuine and loyal champion of water quality in the Valley and encouraged the other directors to fund such efforts regularly.

After some discussion, everyone agreed that we should offer to help the groups tell their stories. After seventeen years and hundreds of grants to a variety of organizations, this was the first time we had discussed the idea of a public relations strategy in detail. We were not particularly interested in getting publicity for VEE—it was created in a blizzard of national and statewide news stories about the Kepone disaster. During our first few years, we spent money on carefully chosen priority areas such as water quality, law and public policy, mediation, and local environmental education projects. Funding the work, rather than communicating about it, was as much publicity as we thought necessary. But this discussion opened our eyes to how few people knew what the groups that we were helping did for the benefit of so many Virginians.

The discussion also illustrated how bringing on new board members inevitably led to innovations in the way VEE operated, as new areas of knowledge and experience were incorporated into the board's mélange. Longtime chair Dr. Dixon Butler, commenting on this consistency, told me: "We always seem to have had the right mix of talent and knowledge on the board when we needed it."

When everyone seemed to have said all they needed to say about helping grantees tell their stories, Dixon Butler appointed a committee to interview some public relations firms and come up with a recommendation. That job fell to Al Smith, Patricia Kluge, and me. In the following weeks, we developed an invitation to perform this work and sent it to three firms.

We interviewed two Virginia firms first. Each was professional, eager, and competent. We asked what they knew about environmental matters, our grantees, and even about us. Both conceded they had not represented environmental stories to this point but were confident that their techniques would translate seamlessly to environmental public relations.

Al Smith and I met with Howard Rubenstein of Rubenstein & Associates one morning at the VEE office at the James Center in downtown Richmond. Howard had flown in that morning from New York City and was dressed conservatively in a navy suit, white shirt, and dark tie.

Al Smith, a veteran politician, was used to people making pitches to get his support. We had listened the day before to the other firms seeking the business, and both of us were curious to see what Howard Rubenstein would be like.

Howard had the highest reputation in the field of public relations, having served as an advisor to a series of New York mayors, the city's business elite, and the Cardinal Archbishop of New York. He had started out as a reporter and moved to public relations after a few years. He had been practicing his craft for decades. I had high expectations for him and was genuinely interested in knowing why he found our invitation attractive. I have always loved a good story and was ready to see how this world-class public relations professional would approach us.

After the usual five minutes of traditional welcome-to-Virginia get-to-know-you chitchat, we sat down at the conference table. He opened his briefcase, which contained several of our most recent annual reports. He proceeded to lay out his understanding of what we did, what we had accomplished to date, and how the Endowment's priorities complemented each other. The confidence and competence he radiated was compelling. Here was someone who two weeks earlier had never heard of the Virginia Environmental Endowment, and yet, at our first meeting he displayed such a thorough grasp of our unique origin and boundless opportunities that he had already developed a public relations strategy to leverage our grants into public support for our grantee organizations. Before we could ask a single question, he had thoroughly summed up our public relations situation as he saw it and suggested a plan for what to do about it. It was stunning how well prepared he was and how thoroughly he "got" what we were trying to do!

"'Foundation Makes Grants' is not news," he said, panning his hands out in a mock headline. "That's what you're supposed to do. 'Friends of the Shenandoah River Launch Water Monitoring Program,' now that's a headline. You don't need to issue press releases; your *grantees* need to do that. That's my first recommendation."

He went on to explain that if we hired his firm, they would meet with our grantees and help them to tell their stories and get more press coverage, with the aim of becoming better known and raising more money to support their work. This is not as easy as it sounds. Nonprofits would often tell me they had enough on their plates just trying to do their job and had no time for "extras" like public relations. Nor were foundations rushing to help them with public relations or general support grants. So, most grantees had neither the time nor the money for public relations and marketing.

Peter Drucker, the legendary management expert, after writing many books about managing for-profit corporations, eventually wrote one called *Managing the Nonprofit Organization,* which became an instant classic. In it, Drucker points out that market research is essential for nonprofits, because they need to find out exactly who needs the kind of services they purport to provide, to delineate exactly what services their group can most usefully deliver to those clients, and to let people know about—that is, publicize—their mission and capabilities.

At the heart of it all is learning how and to whom to tell your story. That is why we hired Howard Rubenstein.

Soon after, Lloyd Kaplan, a Rubenstein senior vice president, called me: "Howard has assigned me to work with you. When can I come down and meet some people?" Over the next few months, Lloyd met with several grantees to explain how he might help them tell their stories. For several of them, he wrote press releases and explained how to reach out to reporters and hook them with a story line. Virginia Conservation Network was helpful in organizing these meetings.

Soon, press coverage of environmental stories picked up noticeably. Rubenstein & Associates tracked environmental stories in the media throughout Virginia. The *Richmond Times-Dispatch,* Virginia's statewide paper of record, assigned a reporter to work full-time on environmental stories. Papers in Norfolk and Roanoke increased their coverage as well. Clean air and water were now being covered more often and in more depth. And many of the first calls these reporters made were to our grantee organizations.

A good example of how this works came from the Valley Conservation Council's Faye Cooper. In December 1996 she sent me a note about our

"Virginia Waters," by Gary Brookins for the *Richmond Times-Dispatch,* June 2008. (Reproduced by permission)

recent grant saying, "Jerry, this made the front page! This type of publicity works wonders—thank you!" The headline read: "Environmental Foundation Awards $50,000 Grant to the Valley Conservation Council."

After Lloyd Kaplan left for another position elsewhere, Cheri Fein from Rubenstein & Associates also helped our grantees gain coverage of their work. Media coverage of the environment in Virginia has been a regular feature of the news ever since. VEE is not responsible for that, but we did give it a boost.

By supporting our grantees' ability to tell their stories, we hoped to increase their impact and help them raise more money. VEE also benefited because our grant funds, already leveraged by the matching funds we required our grantees to raise, would go further and work harder if the nonprofits were to attract even more support.

# 17

# Attitudes of Virginians about the Environment

## THE FIRST POLL

ANOTHER WAY that the Endowment had an influence on public policy was its willingness to conduct public opinion surveys to help measure support for environmental improvement. Our first attempt was with The Nature Conservancy in 1992, when VEE made a grant to conduct a poll to see how much support there was for a parks and recreation bond issue. The poll showed that there was a lot of support, and in the November election, the parks bond won with over 67 percent of the vote.

Virginia was a moderately conservative state in the mid-1990s. 1994 saw a turn to more conservative politicians being elected in Virginia and in the US Congress. Virginians elected George F. Allen as the Commonwealth's sixty-seventh governor. While neither he nor any other politician would ever say that they were against clean air or water, they were sympathetic to industry's complaints about "burdensome, costly, job-killing" regulations. This mantra began in the 1970s and persists to this day, despite unprecedented economic progress in the nation's GNP.

Economics is not the only measure of value, but it is crucial. One of the first books I read about economic policy was written by Dr. Edwin G. Dolan, entitled *TANSTAAFL* (which stands for "There Ain't No Such Thing As A Free Lunch").[1] First published in 1971, it has achieved the status of a classic. Dr. Dolan states in the preface that he wrote the book after observing "that a distressingly large number of writers responsible for bringing the environmental crisis to the center of public attention were as ignorant of the most elementary principles of economics as economists were of ecology."[2] In its most basic application, the TANSTAAFL principle is a simple statement that everything of value has a cost. It is a small volume addressed to laypeople, students, and professionals "in the hope that it will help channel thought and discussion toward the development of a genuine science of ecological economics."

One expert who has made a difference, though, is Dr. Mike Ellerbrock, a professor at Virginia Tech and director of its Center for Economic Education, whose field is agricultural and natural resources economics. A scholar and a teacher, he is also a dynamic speaker who has that rare capacity to make economics comprehensible to those of us who never took a course in it. Further, he has brought his unique talents to good effect by serving as cofounder of the Virginia Natural Resources Leadership Institute (VNRLI), a project jointly founded by the University of Virginia and Virginia Tech. VNRLI is a professional development program for leaders confronted with Virginia's most pressing natural resources issues who seek new skills in conflict resolution and collaborative problem-solving.

The 1994 midterm congressional elections brought a Republican majority to the House of Representatives, and Newt Gingrich, a representative from Georgia, became Speaker. He initiated a program he called his Contract with America, a plan that among other things would roll back the environmental regulations that in his view were stifling the nation's economy. As opposed to the bipartisan nature of political support for the environment in the 1970s, political divides were now becoming apparent.

At the Endowment board meeting in March 1995, we were talking about Speaker Gingrich and his "contract." We didn't see any evidence that people found environmental rules burdensome; in fact, our experience and interactions with both environmentalists and businesspeople said just the opposite. Things could always be better, but we detected no great call for deregulation. In that year's annual report we wrote, "There is a school of thought that suggests that rules and regulations to protect the environment may be too restrictive, and that change is needed to provide redress and unshackle business. Our experience suggests just the opposite: people are working together in large numbers with greater effect to protect the air, land, and water from pollution, impairment, or destruction . . . and expect all of us to do our part. We remain convinced that economic prosperity and environmental integrity are not antithetical, but inseparable."

We did wonder, though, what Virginians thought about environmental rules. Concerned that there was a disconnect between what we saw people doing and what some politicians were saying, we wondered if anyone had ever polled Virginians on their attitudes about the environment. Al Smith, our retired lawmaker and by then an ardent "environmentalist," said, "I've never seen one, but I know the political parties and candidates do polls all the time. I don't think anyone has ever done a survey about the environment or released it publicly if they did." We agreed this poll idea was something worth considering.

Along with Al Smith, kicking the idea around the table were two former Virginia first ladies, Jinks Holton and Jeannie Baliles, and the politically connected Patricia Kluge, all experienced with the pluses and minuses of polling the electorate. We had our sense of what people thought and we had a realistic perspective about polling, but we still felt that we did not have a clear enough picture. In the end, the board's response to the idea was, "Let's find out."

After the meeting, I called Howard Rubenstein and told him what we had discussed. He said, "You must be nonpartisan, or at least bipartisan about this, and to maintain your credibility you can't do a 'push poll.' You can't ask questions to get you the answers you want. You must be more rigorous."

"You have to be willing to accept the results no matter which way they come out," he went on to say. That made sense. Howard provided the information to help us get started: "Call Dick Morris, he has experience polling for Democrats and Republicans. He's been doing polling for businesses and for Republican politicians since 1982 and is an expert at designing useful polls, not these push polls that deliver whatever results you want."

Howard told us that among Morris's former clients was one prominent Democrat, President Clinton, for whom he had done polling when the president was governor of Arkansas. Such a connection might help the cause of environmental protection nationally if the poll results turned out in favor of the environment. And having a pollster who had worked for both political parties would add to its credibility.

We decided to go ahead, and deal with the results later. Thus, in April of 1995 we commissioned our poll of Virginians' attitudes toward the environment. We hired Dick Morris to design and conduct it.

We invited him to Richmond to meet with me, Al Smith, Jeannie Baliles, and, to represent our grantees, Terri Cofer Beirne, executive director of the Virginia Conservation Network. We made it clear to Dick that we wanted a thorough and honest poll that would not be weighted in one direction or the other, whose results we intended to publish for all to see.

It took a couple more meetings and several weeks for the poll to be constructed. I had never helped to design a poll before, and it was a fascinating exercise. To get to as much truth as practical, I learned, one must ask the same question in different ways and with varying specificity, sometimes to elicit a yes-or-no answer, sometimes to force choices between two "societal goods" such as environment or economic development, and sometimes to elicit more nuanced preferences, such as "Yes, economic development, but without environmental degradation."

We read through and discussed page after page of questions, carefully considering each individual question, and were pleased at how Dick Morris had developed not just a range of potential answers ("strongly favor" to "strongly disagree," for example) but also forced choices between environmental and economic activities. Further, to ensure reliability among answers, the full poll revisited some of the questions in different ways at different points and in different contexts. Creating a poll that can reliably demonstrate what people really think is not a simple matter. This kind of poll takes several minutes of conversation with each respondent, not asking a brief question or two.

Dick pressed us about the topics we wanted clear answers to. Environmental regulations—whether to keep them the same, lessen, or increase—certainly were important given the rhetoric of supposedly "costly, burdensome, job-killing" regulations coming out of Washington. We sincerely wanted to learn what Virginians thought about Virginia's environmental issues. Finally, we were all satisfied with the quality of the survey and asked him to proceed. Then we waited, a bit anxiously, not at all sure how the poll would turn out.

The survey interviewed one thousand likely Virginia voters chosen according to a statistically valid sample of the entire Commonwealth. The sample is accurate to a 90 percent likelihood, with a margin of error of plus or minus 2.4 points. The interviews were done on the evenings of May 18, 19, 20, and 21, 1995.

To quote from the report Dick prepared for us, "Voters were asked in different ways whether they think there is too much regulation, or whether regulations should be cut back or reduced. At first, respondents indicated that there should be less government regulation of the environment (41–55). However, when voters are asked to assess whether there should be less regulation by government of each area of the environment, their opposition to decreased regulation grows. Voters reject less regulation of toxic and hazardous wastes by 17–76, and of water pollution by 26–71, drinking water by 23–71, wetlands and coastal areas by 24–63, air pollution by 35–61, and of historic sites by 35–59."[3]

To summarize, in the areas related to water pollution—toxic wastes, water pollution, and drinking water—there was great public animosity to proposals to reduce government regulation. The more specific the potential threat, the greater the support for retaining the protective canopy of regulation. The poll's clearest message was that Virginians wanted to plan for economic development that would *not* endanger the environment—88 percent said so.

These results were more favorable than we had hoped for. Large majorities were against weakening regulations or cutting them back. The results showed that the environment was a really important issue in Virginia. And that was across the spectrum, no matter how you sliced it demographically, geographically, or politically. "Even those who favor less regulation of the environment indicate a preference for reform of regulation over cutbacks," Morris noted. He told me it was the most consistent result he'd ever seen—that the environment was a big issue, even in the predominantly conservative state of Virginia.

On Thursday, June 8, 1995, the Endowment held two briefings to release the results of our poll. The first featured Dick Morris explaining the results to about fifty representatives of the environmental conservation groups from around Virginia. The results were a welcome validation of their efforts, showing that the Virginia electorate strongly supported government regulation to ensure pure water, clean air, and conservation of natural resources.

The first poll's message was a clarion call for preventing pollution in the first place, conserving natural resources over the long term, and protecting the environment for future generations. The poll's overwhelming support for environmental groups' efforts to protect the environment was an enormous boost to their confidence. The poll also showed that the framers of the 1971 Virginia Constitution knew what they were doing when they included Article XI, the conservation article.

Immediately after the briefing for the environmental representatives, we held a press conference to disclose the results to the media. Reporters came from Richmond papers and television and radio stations, as well as from Norfolk and Washington.

I was standing at the back of a large room in the Berkeley Hotel in downtown Richmond anticipating the reaction of the press as Dick Morris energetically presented the poll's findings. He is a short man, but his lively and comprehensive explanation of the poll results made him the biggest presence in the room. "Virginians want a clean environment," he said. "There is no doubt about that."

When he finished his formal presentation, he answered every question posed for the next hour. Reporters were genuinely interested; this was, after all, the first time a poll of Virginians' attitudes about the environment had ever been released. The poll results were front-page news the next morning all around Virginia.[4] Editorial support soon followed.[5]

After everyone had gone, and it was just Dick and I left to relish the coverage this story would likely receive, he said, "Well, Jerry, this has been great fun and we all learned a lot from this survey, but this is my last private job for a while." I asked him what he was going to do. "Work for the president," he told me. "THE president?" I said, surprised. "Yes, on Monday morning I will be a special assistant to the President of the United States, and you better believe he is going to hear about this poll."

As I wrote in an op-ed piece a few weeks later, the results of the poll begged the question: "What about letting businesses reduce pollution in their own way, with minimum standards?"[6] Virginia voters understood that voluntary pollution control was unlikely to become a high priority in the free market system. This is not an antibusiness observation but a pragmatic and realistic one. Although, as noted elsewhere in this book, plenty of businesses have figured out how to protect the environment and save money.

We found—and subsequent polls we published in 1997, by Peter Hart and Glen Bolger, and in 1999, by Dr. Robert Holsworth, confirmed—that Virginia voters across the political spectrum agree that clean water is a top priority. It is a mainstream value, and they expect the government to uphold it.

That first poll influenced public policy on the environment both in Virginia and nationally. VEE's poll was cited in stories in the *New York Times* and *Newsweek* as having an influence on the president's environmental sensitivity and policies.[7] Dick Morris did indeed convince President Clinton to start talking about the environment during his 1996 reelection campaign. Many of the president's subsequent speeches added the phrase "and protect the environment" to whatever subject he was discussing. The big takeaway from the Dick Morris poll, the result that impressed President Clinton, was that "Virginia voters care about the environment!" As a reliably red state at the time, this was valuable political intelligence.

Years later, reflecting on the Endowment's first quarter-century of grant-making, VEE board chair Dixon Butler stated, "That poll is the most important thing we ever did." However, election results in Virginia for the next two years showed much lower rates of people saying the environment was their most important issue.

In 1999 we did another poll. This time we wanted the group to focus on why there was such a discrepancy between poll response and actual voting results. In this poll they found that a strong environmental ethic exists—people are strong for water quality and slightly less strong for other environmental topics—but when they tried to tease out the discrepancy between what people say and how they vote, they concluded that the answer, while not entirely definitive, seems to be that people view the environment as a local matter. Our experience with so many grantees wanting to conserve and protect their sense of place is consistent with this observation. As a statewide issue, the environment doesn't translate well into voting patterns, because people mostly care about their own river, their own backyard. For the most part people in Fairfax don't pay attention to what's going on in the Clinch River—and vice versa; and people in Hampton Roads don't follow the issues that matter to people in Winchester. However, the people who live in those places care very much about what's happening in their own backyard.

Dr. Robert Holsworth, former dean of the L. Douglas Wilder School of Government and Public Affairs at Virginia Commonwealth University, who was our pollster this time, said it comes down to the fact that, particularly in races for the House of Delegates and even to some extent the State Senate, the smaller the district, the more likely the environment is to resonate with voters. With bigger districts and congressional and other statewide races, there are so many other issues people care about that, while they say protecting the environment is important, on a statewide level that importance drops significantly compared to other issues. Since then, as we have learned through subsequent polls led by Quentin Kidd at Christopher Newport University, the public has continued to show a local emphasis in its consistent support for clean air, clean water, clean development.

# 18

# Follow the Money

FISCAL ANALYSIS AND THE 2 PERCENT SOLUTION

THE FIRST THING to know about public policy is that the budget is the real policy document. If you want to see what a government's priorities are, follow the money. Virginia's constitution mandates conservation and protection of the environment in Article XI, but it does not, unlike with Article VIII on public education, mandate money to carry it out. In fact, the Commonwealth's financial support has varied widely from one biennial budget to another for environmental priorities such as water quality, land conservation, and cost-share programs for farmers to reduce poison runoff.

An important, and continuing, source of funds for clean water is the Virginia Water Quality Improvement Act of 1997, sponsored by Delegate W. Tayloe Murphy Jr. It was enacted by the Virginia General Assembly in response to the need to finance the nutrient reduction strategies being developed for the Chesapeake Bay and its tributaries. The act directs DEQ to assist local governments and individuals in reducing point source nutrient loads to the Chesapeake Bay, with technical and financial assistance made available through grants provided from the Virginia Water Quality Improvement Fund. While this kind of commitment by the Commonwealth is necessary, the amount of money provided each year varies with the level of budget surplus.

On November 26, 2001, I wrote Governor-Elect Mark Warner to encourage him to increase funding for the environment: "If you want to make progress in natural resources, then the state's financial commitment for carrying out its responsibilities under Article XI must rise. I suggest you establish a goal of raising the natural resources budget to 2% of the General Fund by the end of your term."

A few weeks later on February 6, 2002, in a crowded room filled with Garden Club women from throughout Virginia attending their annual Legislative Forum, I raised the idea of "putting your two cents in," hoping to start a "two cents campaign," to popularize the goal of appropriating more money for the environment, 2 percent of the state's General Fund.

149

Governor Warner had just been sworn in and there was hope that he would be a champion for the environment. While listing some important environmental priorities that the Garden Club could support, I pointed out that barely 1 percent of the state budget—one penny out of each dollar—was allocated to protect the environment.

I said, "When it comes to the constitutional requirement to conserve and protect our atmosphere, lands, waters, and other natural resources, our minimum goal by the end of the next four years ought to be 2 percent. . . . We need a financial commitment to match the good words about land conservation, water quality, fisheries management, and the goals of the Chesapeake Bay 2000 Agreement . . . and a predictable source of income to sustain it."

I asked them to help restore a General Fund level of 2 percent, a percentage for environmental spending that had not been seen since the Holton administration during 1970 through 1974—and to let the General Assembly know that it is their duty to appropriate sufficient money to carry out the mandate in Article XI.

At VEE we suspected that state financial support for the environment was not what it could be, but we needed to document exact numbers. Through a grant to the League of Conservation Voters Education Fund, we commissioned a respected budget analyst, Jim Regimbal of Fiscal Analytics, LLC, to not just study the Virginia budget but also review other states' allocations for natural resource and environmental protection.

The results were sobering. His report stated, "Natural resource funding has decreased significantly and is going to drop more. General Funds in the 2004 budget for Natural Resource agencies are now at a relative level not seen since the mid-1980s with only about six-tenths of a percent (0.6%) or $76 million of total general funds available for appropriations in the adopted budget."[1] And with respect to the Commonwealth, the report continued, "Virginia's level of natural resource funding is low relative to surrounding states and national averages. . . . The US Census Bureau ranks Virginia dead last in state natural resource and parks spending." Despite our years of effort, Virginia's situation was equally dire on funding for land conservation: "With specific regard to land conservation, a number of eastern seaboard states including Delaware, Florida, Maine, Maryland, New Jersey, North Carolina, Pennsylvania, and South Carolina all have various forms of dedicated funding sources for land conservation."

The VEE board of directors was appalled at this situation. We sent the report to the press, the governor, and the legislature as well as to

environmental groups. Most were shocked by Virginia's "dead last" rank-
ing. Getting somebody to do something, however, was another matter.

Tayloe Murphy, now Governor Warner's Secretary of Natural Resources,
speaking at the Environment Virginia conference on April 10, 2002, in his
graceful and generous way presented the Endowment with a certificate of
recognition signed by Governor Warner, praising Virginia Environmental
Endowment's twenty-five years of grant-making to improve the environment of Virginia. In accepting the recognition on behalf of the Endowment, I made a point of recognizing in turn the hundreds of groups we
had helped, because "they're the ones who actually do the work."

Addressing the matter of money, I continued: "As Tayloe knows better
than anyone in this room, budget is policy in the real world; and when you
are putting less than a penny out of each tax dollar into natural resources
and environmental protection, you have devalued the only world we have.
I want to commend Tayloe for stepping up to that responsibility. I hope we
can get to two cents by the time Governor Warner's term concludes. . . . I
urge you all to get behind that, because only with that type of public investment will there truly be a balance that is critical for a healthy economy
and a healthy environment." Later that year, I reiterated the same call to
the annual meeting of the Virginia Conservation Network on October 5,
2002, in Williamsburg and commended Secretary Murphy for embracing
the 2 percent goal.

In the spring of 2003, Secretary Murphy convened the first-ever Governor's Natural Resources Leadership Summit. This two-day event in
Williamsburg attracted over three hundred people from business, government, academic, and nonprofit organizations. The group separated into
several small groups, each exploring different aspects of natural resources
policies. The result was a prioritized list of ideas to improve various parts
of the environment, and another call to fully fund these areas. Secretary
Murphy had led the group in creating a specific environmental agenda for
the governor.

At the December 3, 2003, Outreach Award Dinner of the Virginia
Manufacturers Association (VMA), I reiterated the need for reaching the
2 percent goal for natural resources funding: "The gap between this core
constitutional responsibility and the amount budgeted to do the job is a
disconnect that has produced the sorry result of Virginia's being ranked
last among the states in financial support for the environment."

That fall, through the efforts of The Nature Conservancy, a steering
committee of business and conservation leaders had been formed to

mobilize and plan a campaign to convince the governor and the legislature to increase natural resources funding—and put its money where its constitutional mandate was. I asked the VMA to join the coalition and help persuade government leaders of the merits of this idea. Indeed, several of their member organizations got involved.

In 2003 we invested $93,450 in The Nature Conservancy, which had volunteered under the leadership of Michael Lipford, who had become the Virginia Nature Conservancy's state director, to help to build a formal coalition of business and conservation whose only agenda would be to lobby the governor and the legislature to increase funding for natural resources programs. Lipford was skilled at developing environmental information programs and collaborations with businesses. His polling and public relations efforts during the 1992 and 2002 referendums on whether to issue state bonds for new state parks and recreation areas had helped those measures pass by a two-thirds majority. He was the perfect person to build such a coalition. We knew he would be effective in circulating the message that adequate funding for the environment was in both groups' interests, because degradation of the environment is bad for business and has a negative impact on the economy.

In 2005 we added another $90,000 to the effort. Michael by now had recruited some of the conservation organizations in Virginia to work together with business groups that had been recruited by Dennis Treacy to work together on the funding increases. Dennis previously served as the Director of the Department of Environmental Quality and was now serving as executive vice president and chief environmental officer of Smithfield Foods. Over the next year, VEE invested more funds to help this new group to become an independent corporation, whose mission was to increase general fund appropriations for water quality and land conservation. They named it VIRGINIAforever.

In 2006 we made a grant of $110,125 for transitioning the campaign from under The Nature Conservancy to VIRGINIAforever. Nikki Rovner of The Nature Conservancy succeeded Michael Lipford in coordinating this effort. VIRGINIAforever was formed with equal representation on its board of directors from environmental nonprofits and leading businesses and law firms. VIRGINIAforever was and remains a unique partnership. We continued to support this new entity for the next two years with $50,000 more.

Meanwhile, in another effort to protect conservation lands in perpetuity, Governor Warner and, later, Governor Kaine each established four-year objectives to protect 400,000 acres of Virginia land during their terms, which they succeeded in carrying out. Many of VEE's grantees helped them achieve those objectives.

The problem that remains is how to sustain and institutionalize these results so that the funding becomes more consistent. Having a capital spending plan to invest those funds in land conservation and water quality would help. Such a plan would spell out exactly the costs and timetables of projects to be funded each year. VIRGINIAforever is working on that too. One reason the park and recreation bond issues passed so handily is that they contained lists of exactly what projects would be supported in localities across the Commonwealth.

The Water Quality Improvement Fund was amended in 2005 to reflect up-to-date water quality needs and priorities in Virginia, particularly the issues of implementing Chesapeake Bay "Tributary Strategy Plans" and removing Virginia waters from the Clean Water Act list of impaired waters.[2] Annual appropriations to this fund from fiscal years 1998 through 2017 totaled $908.28 million but have varied from a low of $10 million in 1998 to a high of $197.33 million in 2007, with great variances from year to year. The total includes three years of bond proceeds, totaling $415 million. During the fiscal years ending June 30, 2021, and June 30, 2022, Virginia's budget surpluses totaled $2.6 billion and $1.94 billion, respectively.

Slowly but surely, the natural resources funding levels are rising. Follow the money. The leverage of VEE's investments in the 2 percent solution has been outstanding. Thanks to VIRGINIAforever's efforts to focus attention on natural resources funding, hundreds of millions of new state dollars were appropriated for land conservation and water quality improvements in recent years.

# 19

# Environmental Education

PROMOTING ENVIRONMENTAL KNOWLEDGE

The world is full of magic things, patiently waiting for our senses to grow sharper.

—William Butler Yeats

VEE INVESTED more than a million dollars in environmental education during my tenure. Not every grant the Endowment made was for tens of thousands of dollars. We saw that small, targeted grants for local environmental programs could also have a big impact. Incorporating nature into learning to read, write, count numbers, and play is a powerful motivator.

The year 1981 saw the first of several grants to the Richmond regional Mathematics and Science Center. The Center provided in-service teacher training in the summer to high school biology teachers. The training was designed to be used by the teachers to prepare their students for their field trips during the following school year. The classroom instruction included information on toxic substances and their effects on water quality and marsh ecology. The highlight, of course, was the field trip itself. This allowed students and teachers to learn by way of canoeing and hiking wetland environments in a variety of Virginia localities. In an academic year, more than two thousand students from public and private schools around the metro Richmond area would take part. Later, in the mid-1990s, we also funded the popular Teachers on the Bay program.

At about the same time, there was a local group in Virginia Beach led by George Hagerman. He was a retired Navy captain, a US Naval Academy graduate, and a World War II veteran who was now devoting his retirement years to protecting the environment and educating people about why that is so important. He had the time, energy, and organizational ability to get things done. He named the group Citizens Program for the Chesapeake Bay (CPCB).

We made our first grant to CPCB to establish a mini-grant program. Its purpose was to fund small local projects that would improve the quality of the Chesapeake Bay by focusing on its local tributaries. Much was made in the 1990s by the federal-state Bay Program lauding its new "tributaries strategy," but George Hagerman had already had that idea in the late 1970s. He was truly a man ahead of his time, and a great, lively presence wherever he went. He had a ruddy face reflecting how much time he spent outdoors, and a ready smile, even as he faced daunting odds. He insisted that local buy-in was necessary for success and persisted in launching a program of educational projects that made a big difference to local communities.

With a lot of volunteer time and enthusiasm, many projects were successfully carried out—without much money. In 1980 we awarded $25,000 to George and his group, and they in turn distributed it in small, local grants of about $2,000 each. This mini-grant program enabled local groups to stimulate cooperation among businesses and local governments interested in making environmental improvements in their own communities. In the 1970s there was a saying: think globally, but act locally. They were truly acting locally, translating national policies into local action.

We awarded CPCB another $25,000 the following year, and a third grant of $35,000 in the next year, all to help improve the quality of the Chesapeake Bay and its tributaries. Together these grants in turn funded twenty-one different projects benefiting thousands of people in the Richmond, Tidewater, and Northern Neck regions. They also greatly improved public awareness of environmental matters.

Based on CPCB's success, we partnered with the Virginia Foundation for the Humanities in 1982 to extend the mini-grant program statewide, expanding its focus into land and community issues as well as water quality. We also made our first grant for environmental education to the Chesapeake Bay Foundation, a catalytic grant that led to major financial investments in CBF's educational programs over the next several decades by other foundations and by the Commonwealth of Virginia.

## Mary Baldwin College and Lewis Creek Cleanup

One example of how students, colleges, and communities can cooperate comes from Staunton, Virginia.

In the spring of 1982, John Deehan, executive director of SURE (Staunton United Revitalization Effort, Inc.), a civic organization, brought to the attention of Mary Baldwin College (now University) in Staunton the foul odor of Lewis Creek, a waterway which flows through the heart of Staunton. SURE, which served as a promoter of businesses in Staunton, was concerned that the rotten-egg odor from the Creek would discourage customers from shopping in the city's downtown. Mr. Deehan asked the college for help in investigating the reason for the creek's offensive smell.

In response to this request, microbial tests of the creek were carried out in the summer of 1982 by summer science students and in the fall of 1982 by a microbiology class at the college under the direction of Dr. Lundy Pentz, assistant professor of biology. After taking this microbiology class, a student named Joanna Campbell conducted chemical and microbial testing of Lewis Creek in the winter and spring of 1983 under the supervision of Dr. Pentz and submitted her report to Mr. Deehan in May 1983. Mr. Deehan sent the report to Gene McCombs, Staunton's city manager. Campbell's report provided supplemental data supporting sewage being a major contaminant in Lewis Creek.

Joanna Campbell then decided to pursue the study of Lewis Creek for her senior project and make use of both parts of her double major in economics and biology. The first part of her project would consist of additional data collection and analysis carried out under the direction of Dr. Pentz. The second part of her project would consist of conducting a benefit/cost assessment study of Lewis Creek for the City of Staunton under the supervision of Dr. David Molnar, assistant professor of economics at Mary Baldwin. However, funding was needed for the equipment to complete this project. During the summer of 1983, Campbell researched possible sources of funding for her senior project and was referred to the Endowment. She wrote a proposal requesting funding for the project, and it was submitted to VEE by the college in October 1983.

The Endowment awarded Mary Baldwin College a grant of $2,875 in November 1983. Following the announcement of this grant, Joanna Campbell, Dr. Pentz, and other representatives from the college met with Staunton city mayor Dolores Lescure and city manager McCombs. Both were enthusiastic about the grant award. Mayor Lescure had long been a proponent of cleaning up Lewis Creek, and her advocacy and tenure on the city council were critical to the success of the VEE grant. Ms. Lescure served on Staunton's city council from 1980 until 1987 and was chosen by the council as mayor in 1983 and vice mayor in 1986. Prior to serving on Staunton's city council, Ms. Lescure worked at Mary Baldwin in a variety of positions, including as a teacher of journalism.

Joanna Campbell completed her final report on Lewis Creek in June 1984. The report included microbial test results; counts for total coliform bacteria, which can be found in soil, plants, and the intestines of warm-blooded animals; and counts for total fecal coliform bacteria, which can only be found in the intestines of warm-blooded animals. Campbell was able to distinguish between total coliform bacteria counts and fecal coliform bacteria counts because of the college's purchase of a special incubator with grant monies provided by VEE. Given that a dramatic rise in fecal coliform levels, which indicated sewage contamination, was observed at two sampling locations, statistics were applied to evaluate the difference in these fecal coliform levels. As a result of this analysis, it was determined that the rise in fecal coliform bacteria levels between the two sampling locations was significant, at the 95 percent confidence level, thus pointing to an influx of sewage contamination between these two locations.

The report's benefit/cost assessment described some of the benefits associated with replacing the relevant sewerage line between these two sampling locations and provided a cost estimate. For comparison purposes, the cost estimate was presented as a percentage of the fees collected by the City from residents for use of the its sewer system over the course of one year.

In July, Joanna Campbell submitted her report to Councilor Lescure (her term as mayor had expired on June 30) and city manager McCombs. Then, on October 7, 1984, the local paper in Staunton ran an article on Lewis Creek and the sanitary sewer line identified in Campbell's report. The article stated that the City had investigated the sewerage line and discovered that it was partially collapsed and leaking into the creek. Further, the City had received bids on replacing the damaged 646-foot terra-cotta line with a concrete pipe. A City engineer, who was quoted in the article, thought installing this concrete pipe, along with the replacement of about 136 feet of another sanitary sewer line (which, the City determined, overflows when there is a heavy rain), should rid the creek of most of its foul smell. He agreed with Campbell that although these projects would not make Lewis Creek a pristine waterway, they would be helpful.

In August 1986, vice mayor Lescure and Mr. Deehan presented Joanna Campbell with an award "For Outstanding Citizenship and Dedicated Community Service" on behalf of the Staunton city council. Also demonstrating appreciation of Campbell's work with Lewis Creek, a resolution was passed by the city council. Following the completion of her graduate work at the University of Virginia, Joanna Campbell went on to a long and rewarding career in public service at the US Environmental Protection Agency in Research Triangle Park, North Carolina.

VEE's funding of the Lewis Creek project continues to provide benefits to individuals and the Staunton community. For example, several upscale restaurants with outdoor seating are now comfortably located close to the location of the problematic leak that was identified in the 1984 report. Additionally, Staunton's city council formed the Lewis Creek Watershed Advisory Committee in September 2004, with one aspect of its mission being to conserve and actively promote the improvement of the waters of Lewis Creek. Until his retirement in 2014, Dr. Pentz and his students continued to monitor Lewis Creek annually. During the ten years after the formation of the Lewis Creek Watershed Advisory Committee, Dr. Pentz's students submitted their findings through the Committee to Staunton's city council. The monitoring of Lewis Creek has led to a marked improvement in the creek and greatly enhanced public awareness of the creek's importance.

Just a small grant leveraged measurable improvements to the quality of water of Lewis Creek, the quality of downtown Staunton, and to ongoing benefits of a collaboration between a civic organization, a city, and a local college, and made a lasting difference to a community.[1] As Joanna Campbell put it, "I think I was an important resource for the community in the sense that I changed people's perceptions about what one person can do."

Dr. Lundy Pentz, now emeritus associate professor of biology at Mary Baldwin University, told me:

Joanna Campbell's involvement with Lewis Creek in Staunton, Virginia provides a striking and dramatic [demonstration] of the practical application of microbiology, chemistry, and statistics by a talented and energetic undergraduate, and the importance of independent funding agencies willing to back small projects like this one, in guiding local government's decision-making. It additionally highlights the power of a double background in science and economics, which permitted Ms. Campbell to make the case for the value of doing the remedial work as a way to overcome the usual resistance to acting. Her work had a profound impact on Staunton, not least because the downtown development has resulted in several upscale restaurants with outdoor seating being located within a block of the obnoxious leak's location. This example provides an important glimpse of what one dedicated and capable person can do by drawing together good science, sound analysis, economics, private foundations' support, and a small college to make a lasting difference in the community.[2]

By 1985 VEE had made environmental education in the service of local improvements an ongoing part of its mission. The return on investments in the mini-grant program was remarkable. People made a difference in their own communities as volunteers.

While building an environmental education program, the board wanted to see what might be done for students interested in environmental careers. Sydney and Frances Lewis, two of the original board members, had stepped down at the end of 1983 and were succeeded by former Virginia first lady Virginia "Jinks" Holton and Dr. Dixon M. Butler. The board wanted to recognize the Lewises for their service to the Endowment and thought some special grant in their honor would be a good way to do so. I was asked to develop this idea into a specific proposal for the board to consider.

Dr. R. Dean Decker, a professor of biology at the University of Richmond, was one of the first people I consulted. I explained what we were trying to do, and with a twinkle in his eye, he suggested that he just might have the answer: encouraging student scientists through the annual Virginia Junior Academy of Sciences symposium for high school students. The Virginia Junior Academy of Sciences (VJAS) has, since its formation by the Virginia Academy of Sciences (VAS) in 1941, initiated and promoted a variety of important programs, including an annual meeting, the presentation of awards, sponsoring radio science quiz shows, and publishing scientific documents. The VJAS has grown to over one hundred affiliated schools. Many of the Virginia Academy members have continued to guide its affairs.

Dr. Decker, an enthusiastic proponent for the VJAS, told me about the annual symposium. Students from around Virginia conduct research in a variety of topics, prepare scientific papers describing the work, and have the opportunity to present them in front of VAS scientists at the symposium. Small cash prizes are awarded when the results are announced, but, as one might imagine, in a large theatre or auditorium filled with hundreds of teenagers, the real reward is the enthusiastic cheering supporting their schools' winners for excellence in science.

The purpose of the VJAS symposium fit well with VEE's desire to encourage student interest in environmental sciences. I asked how we could help.

As often happened when I presented an idea to the board, they would discuss, review, and analyze it from their different perspectives, and by the time we finished talking about it, it would develop into a better idea. So it was with the discussion of how we might help VJAS.

The Lewises had a reputation as generous philanthropists and they like to think big. Considering the nature of the people we were honoring, the board voted to fund an annual $10,000 college scholarship VJAS prize, named for the Lewises, to be awarded at the annual symposium to the student whose paper was named the best by the judges in the environmental category. This relatively large prize (the highest VJAS prize at the time was $100) also stimulated great interest in VJAS and attracted more entries for the competition. It continues to this day and still encourages interest in environmental sciences, our original goal.

The board so liked this idea that when Judge MacKenzie retired, we set up another scholarship in his honor, this one valued at $5,000, emphasizing James River research projects. Since the first scholarship was awarded, VEE has contributed more than $600,000 toward the higher education of outstanding young people.[3]

Another initiative was to develop a partnership with the state Department of Environmental Quality's Office of Environmental Education and with Virginia businesses to provide "outdoor classroom" grants to teachers. Any elementary school teacher will attest that they often must use their own money just to obtain basic classroom supplies. Thanks to the leadership of Dennis Treacy of Smithfield Foods, this partnership made grants of $2,000 to $5,000 each to enable teachers to help their students to construct outdoor classrooms. It was like manna from heaven. Giving grade school teachers $2,000 or $3,000 for an outdoor classroom or for science equipment had an enormous impact on what could be done for the students. Often these grants allowed students to combine learning art, math, and scientific principles in the process of developing and using their new learning environments. And they got to reconnect with nature, an oft-neglected feature of their wired lives.

In 1990 the Endowment provided a grant for the Virginia Association of Environmental Educators to hold the first of what turned into a series of annual conferences, which we also supported.[4]

ADULTS CAN benefit from continuing education as well. In 1991 we initiated the Environment Virginia conference held each year at Virginia Military Institute. This is now an annual event for businesses, government agencies, and environmental groups to gather in a relaxed environment to learn about current issues.

The Environment Virginia conference had its origins in 1990 as a by-product of a 1988 request for $21,144 from VMI professor Ron Erchul to "inventory illegal refuse dump sites in sinkholes and caves in the karst regions of western Virginia, and to recommend strategies for local enforcement and public education" about the risk such dumping posed to groundwater. Retired US Navy Captain Ron Erchul was a professor of civil and environmental engineering.

By this time, VEE had been in business long enough to learn that scientific studies do not always lead to action in response to the results of the research. The Endowment board was willing to fund the proposal but wanted to know, "Is this going to be another academic study that recommends 'more research is needed,' or is something useful going to happen if we fund this?" The board asked me to suggest to Professor Erchul that perhaps VMI would consider holding a conference to discuss the study's results with potentially interested parties. He liked this idea, and VEE funded the research. Toward the end of the grant period, Captain Erchul organized a meeting of interested people to come to VMI and discuss the study's results.

That meeting went so well that Professor Erchul asked the Endowment if it would support another conference the following year. He saw such a conference as an opportunity for people with different perspectives and responsibilities to come together and learn about environmental problems and solutions from each other. We considered this request to be in keeping with one of our core goals, to bring people together to improve the quality of the environment. Ron Erchul shared that goal and wanted to do something practical about it.

A conference with a wider scope and a larger audience quickly became his vision. We funded it a second time, a third, and a fourth and have continued doing so for more than three decades now.

In the years that followed, he added state agencies and private businesses as well as local governments and academics to the list of invitees, recruited and signed up additional sponsors, and expanded the list of topics for discussion. The conference soon became an annual "must do" for a diverse group of participants, who all share the desire to improve the environment and to exchange constructive ideas for doing so.

No other state has such a forum to bring leaders together to build understanding, promote learning, and find common ground in their diversity of approaches to a clean environment. Some of the best conversations take place in the reception for all attendees at the end of the day's program, and later at the bars and restaurants in Lexington where conferees gather to drink beer, eat comfort food, and debate ideas they heard throughout the day. There is nothing quite like sitting at a crowded table at one of these places, with beers and snacks and people squished together, to fire up a lively conversation and let people talk with each other across whatever policy or political divides they might have. The Environment Virginia conference enables those exchanges to happen each year.

A few years ago, during a break in the conference, a woman approached me and introduced herself. She was an environmental professional who has been working in the field since obtaining her degrees at Virginia Tech. She thanked me for a grant VEE had made twenty years ago to a professor at Tech. She had needed a summer job to help her pay her university expenses, and she became a research assistant on the project. She told me how that one job inspired her to pursue her environmental career. She went on to say that there must be many people like her that our grants have helped. It was a kind and gracious thought, and it made me realize that one of the important ripple effects of VEE's grant-making must be the number of environmental professionals' careers we helped to launch along the way.

In 2019 VMI and VEE, sponsors (which by now included state environmental agencies and many businesses too), and attendees celebrated the thirtieth annual Environment Virginia conference. Over the years, every governor of Virginia has addressed the conference at least once, and many leading scientists, public policy experts, and business and nonprofit leaders have brought their unique perspectives to engage participants at the conference. The relationships built and sustained among business, nonprofit, and government people over time is a legacy of this outstanding effort. It has worked out much better than we imagined, thanks to Ron Erchul, the steady support of VMI, and all the enthusiastic participants who make it happen every year. The participants who attend each year—and the relationships they build—help to make the goals of Article XI more achievable year after year.

MEANWHILE, SEVERAL environmental advocacy groups have also developed education programs. Elizabeth River Project has The Learning Barge, and while Valley Conservation Council does land conservation and James River Association does river conservation and protection, both also do environmental education as part of their mission. This is an interesting way of looking at environmental education, tying the growth of environmental organizations to an opportunity for them to provide environmental education as a hands-on experience for children and adults.

The Virginia legislature has from time to time supported the idea of environmental education, beginning in 1974 at the end of the Holton administration. Thanks to the leadership of Delegate James H. Dillard II, a Republican representing Fairfax, it did provide funds to distribute statewide a model environmental education curriculum. Delegate Ken Plum, a recently retired Reston Democrat first elected in 1977, championed environmental matters in the General Assembly for more than forty years. An educator by profession, he understood how important education about the environment was. Plum's support was strong, constant, and effective. While governor, Jim Gilmore established an Environmental Education Commission, which developed the first statewide environmental education plan.

Environmental education has a strong bipartisan history. May it ever be thus.

# 20

# Science Matters

FISHERIES MANAGEMENT IN THE CHESAPEAKE BAY

It is difficult to say how many fish live in the Chesapeake Bay. The Bay is the largest and most biologically diverse estuary in North America, home to 348 species of finfish and 173 species of shellfish, many of which have been fished commercially and recreationally for generations. Its ecological, economic, and cultural importance makes it a resource of worldwide importance. It is an economic and environmental powerhouse.[1]

Demand for seafood and advances in technology have led to fishing practices that are depleting fish and shellfish populations. Until the twentieth century, harvests of many fisheries in the Chesapeake Bay were abundant. Oysters were so plentiful that in the early 1600s Captain John Smith of the Jamestown colony reported that you could practically walk across the James River on the tops of oyster reefs. Today, because of overharvesting, pollution, and disease, oyster harvests are a small fraction of their historic levels.

For most of the twentieth century, the absence of scientific information to guide the establishment of sustainable levels of harvesting of the Chesapeake Bay fisheries resulted in overfishing of several commercially and ecologically important species, such as rockfish, blue crabs, and oysters. Scientific knowledge about the state of the Bay is crucial but must be combined with public policy and regulation for the long-term conservation of the fisheries. And on the intersection of science and public policy there hangs quite a story.

The Chesapeake Bay Foundation sounded the alarm about the many threats to the Bay in the late 1960s, adopting as its motto, "Save the Bay!" In 1982 Virginia Governor Charles Robb appointed a Blue Ribbon Commission to chart the future of Virginia. This commission identified numerous natural and economic threats to the Chesapeake Bay and recommended both state and federal actions to address them, particularly that Virginia fund its own Chesapeake Bay Plan.

In 1983 the first Chesapeake Bay Agreement between the federal government and the states that border the Bay was signed. By 1987, a more comprehensive Agreement was adopted that specifically called for conserving the Bay's fisheries. In the years since the 1987 Agreement, there has been slow but steady progress toward meeting the goals set forth in that document. For example, all parties have agreed to work together toward meeting specific numeric targets for reducing water pollution of the Bay by 2025. However, one of the most difficult remaining problems concerns the overfishing of certain species, such as crabs and rockfish.

For years, fisheries within the Chesapeake Bay were managed on a species-by-species basis, with management plans that did not account for factors such as the abundance of competitors, predators, variations in environmental conditions, and forage species. Granted, the relevant questions—How do you count the fish in the sea? How many fish should be caught each year?—were difficult to answer. The productivity, abundance, and diversity of the fishes in the Bay are among the best measures of its health—not just ecologically but economically, too, because of their commercial and recreational value.

In the 1990s, fishery scientists and managers needed to learn why the numbers of commercially valuable species of fish and shellfish caught in the Bay were declining. The management of Bay fisheries was complicated by the overlapping jurisdictions of the states of Maryland and Virginia, which both border the Bay, and the interstate regulatory body, the Atlantic States Marine Fisheries Commission (ASMFC). The regulatory jurisdictions broke down, roughly, as the states of Virginia and Maryland each regulating species that tend to remain in state waters (everything within three miles of the coast), for example oysters and crabs. For species that stay within three miles of the coast but cross state boundaries, the ASMFC had overall control, and the states had to comply with the plans developed by ASMFC or face penalties. However, each season, when commercial fishing companies asked for the permit to harvest fish, the catch limits imposed were usually based on the level they had caught previously, not on scientific data. Few had questioned this approach until the annual numbers of commercially viable species began to drop.

The decrease in oyster catches had been happening since the early 1970s, and while there was some concern about that among oyster gatherers, whose livelihoods depended on bountiful harvests, and the scientists who were trying to define sustainable levels of oysters, the solutions typically offered amounted to trying to increase the harvest somehow. The idea of limiting the catch in such a way so as to promote the long-term health of the oysters found little support from oyster gatherers because of the limits that action would place on their income—which was declining anyway because of the decline in numbers.

The blue crab harvests varied substantially as well and were threatened by the harvesting of young female crabs before they could reproduce, further limiting the catch. Harvesting so many young females meant that there were fewer of them to generate new crabs, thus exacerbating the decline. At various times in recent decades, Maryland and Virginia have sought to improve the long-term viability of the crab population.[2]

In the year 2000, the new Chesapeake 2000 Agreement made managing the fisheries' harvest levels a priority. The new goal was "to maintain their health and stability and protect the ecosystem as a whole." To accomplish this goal the Agreement called for a new multi-species, ecosystem-based management plan to be developed by 2005.

The Chesapeake 2000 Agreement's call for this new plan was historic and unprecedented. Up to that point, fishing had been done in a largely data-free environment. The Agreement's approach called for a sustainable, evidence-driven management approach. On a coastwide basis, 1982 was the first year that a program was put in place to take fish from the total commercial catch and check what ages were being harvested, which formerly had not really been monitored at all. Mandatory reporting of the catch to the Virginia Marine Resources Commission, the state's coastal fisheries manager, only began in 1994. Before that, there was no system for accountability. The pertinent part of the Chesapeake 2000 Agreement states:

> The health and vitality of the Chesapeake Bay's living resources provide the ultimate indicator of our success in the restoration and protection effort. The Bay's fisheries and the other living resources that sustain them and provide habitat for them are central to the initiatives we undertake in this Agreement. We recognize the interconnectedness of the Bay's living resources and the importance of protecting the entire natural system. Therefore, we commit to identify the essential elements of habitat and environmental quality necessary to support the living resources of the Bay . . . we will manage harvest levels with precaution to maintain their health and stability and protect the ecosystem as a whole. . . . Our actions . . . must be continually monitored, evaluated, and revised to adjust to the dynamic nature and complexities of the Chesapeake Bay and changes in global ecosystems. To advance this ecosystem approach, we will broaden our management perspective from single-system to ecosystem functions

and will expand our protection efforts by shifting from single-species to multi-species management.[3]

To carry out this "ecosystem based, multi-species" approach, the Agreement established the following objectives for multi-species fisheries management:

—By 2004, assess the effects of different population levels of filter feeders such as menhaden, oysters and clams on Bay water quality and habitat.
—By 2005, develop ecosystem-based, multi-species management plans for targeted species.
—By 2007, revise and implement existing fisheries management plans to incorporate ecological, social and economic considerations, multi-species fisheries management and ecosystem approaches.

For the first time, there would be plans to manage fisheries in ways that reflected predator-prey relationships; reproduction levels; pollution from air, land, and water; and habitats such as underwater grasses and shoreline wetlands. There was a consensus among the EPA and the governors, scientists, and environmental activists that this approach was the only way to ensure the long-term health of the Bay.

The question was, how to do it. This new approach also required that the proverbial number of fish in the sea needed to be known more accurately before catch limits could be set. Unfortunately, hardly anyone was sure about how to calculate those numbers in the Chesapeake Bay.

SINCE ITS first round of grants, the Endowment has emphasized the role of science in the development of fact-based environmental policy. Following through on the Endowment's long-standing commitment to help restore the natural resources of the Chesapeake Bay, we were curious to learn how the relevant government agencies were going to develop this new ecosystem and fisheries management plan.

The traditional approach offered little scientific basis upon which to estimate the remaining stock (amount) of any given fish population in the Chesapeake Bay. If one was allowed to catch 50 million Atlantic menhaden last year, for example, were there 50 million left, or 500 million, or 5 million? No one knew. Furthermore, until several years of harvest data were gathered, it would remain difficult to ascertain whether fish stocks were declining, growing, or maintaining, other than by relying on anecdotal evidence from year to year.

In the spring of 2001, the Endowment board was meeting to review grant applications and to discuss priorities going forward. As a former chair of the Chesapeake Bay Program Citizen Advisory Council, I recognized the game-changing nature of the Chesapeake 2000 goal of developing a new fisheries management plan and brought it to the attention of the board. The board quickly picked up on the fact that the objective for having a new management plan by 2005 lacked a plan to achieve it.

Former governor Linwood Holton called for VEE to do something to get this new plan moving. He was an avid enthusiast for the Bay and its oysters, which, as an "oyster gardener," he personally grew off his dock at his Corrotoman River home. He grew about two thousand oysters annually and often said, "I eat them all!" Bob Smith, who for years had been concerned about overfishing of menhaden in the Bay, agreed: "If this might help solve the overfishing problem, I'm all for it." Governor Holton leaned in over the table and said, "Let's find out what it will take to develop this plan and get on with it."

The rest of the board agreed that VEE should investigate this matter, and I was directed to contact knowledgeable people around the Bay and report back with some ideas about how we could help. One of the most productive conversations I had was with Dr. Ed Houde, a professor at the University of Maryland's Center for Environmental Science and Chesapeake Biological Laboratory, a scientist with decades of experience with Bay fisheries. Dr. Houde welcomed the Endowment's interest in learning what needed to be done to move the 2005 ecosystem plan along. He greeted me warmly and was happy to tell me how he saw things. I met with him at his office overlooking the Patuxent River and the Bay, where evidence of his research publications and love for the Bay were on full display.

He told me in his quiet but genuine way that not much was happening yet to move the 2005 plan along. In the biological reality of the Chesapeake Bay, he continued, species interacted with each other in various ways as predators and prey and as competitors for nourishment, while also simultaneously subject to habitat losses and pollution loads. All of this happens dynamically at the same time, he explained, so any approximation of the number of a given species of fish that should be allowed to be caught commercially needs to be made with a better understanding of how many exist in such a rapidly changing environment.[4] He suggested that the Virginia Institute of Marine Science would be a great place to look for help.

Following Dr. Houde's advice, I contacted Dr. Eugene Burreson, director of research at the Virginia Institute of Marine Science (VIMS), a school of the College of William and Mary. VIMS has a three-part mission: to conduct research in coastal ocean and estuarine science, educate students and citizens, and provide advisory services to policy makers, industry, and the public. Chartered in 1940, VIMS is currently among the largest marine research and education centers in the United States. VIMS is also the graduate school of marine science at William and Mary.[5]

Dr. Burreson introduced me to a young postdoctoral research fellow named Dr. Robert Latour, who had earned his PhD in biomathematics. We met in Dr. Latour's "office," which was such a small space that the two of us filled it easily. The room lacked windows, and the only light was from a harsh overhead fluorescent bulb illuminating the reams of scientific papers and documents that spilled from the desk, strained the bookshelves, and piled on the floor, further constricting our movements. We each had a chair to sit in.

The scale of the project didn't fit into existing ones at VIMS. And Rob Latour's background was in math, modeling, and statistics. "Never had a course in marine science," he told me. But he was a good fit for the project because of the need for large-scale data acquisition and analysis to determine how many fish exist at any given time. The timing was great for him, since he would soon reach the end of his postdoctoral term at VIMS. The opportunity to calculate fish populations in the Bay would enable him to remain at VIMS.

Dr. Latour gave me a tutorial on what needed to be done to develop the kind of plan envisioned by the 2000 Agreement. He told me: "Calculating the number of fish is done routinely for the northwest Atlantic Shelf and in other marine areas along the US coast and internationally." Making such calculations in estuaries such as the Bay, he went on to say, is often challenging, because animals come and go seasonally, not to mention that the large-scale data collection programs are often absent. By the end of our meeting, I concluded that VIMS was the place and Rob Latour was the person for developing the new kind of fisheries data and analysis needed to create the management plan called for in the Agreement.

At the fall 2001 VEE board meeting, I presented my recommendation and a proposal on behalf of VIMS. The board members were delighted that we could have a Virginia institution begin this work and moved to support it with $639,000 ($1.07 million in 2022) for three years. This was a large grant for a small foundation with assets of about $21 million, the largest we had ever made. The board was excited about the idea and eager to make the kind of commitment necessary to induce real change.

As Dr. Latour later told me, "The success of those first three years (of the VEE grant) essentially jump-started my career, such that in 2004 I was able to compete successfully for a tenure-eligible assistant professor position at VIMS. Never would have been possible without the program-building that occurred from 2001 to 2004. And now almost twenty years later I'm a full professor with one of the largest research programs at VIMS."[6]

SOON THEREAFTER, Dr. Latour discussed his new project with a VIMS colleague, Chris Bonzek, who had a small grant for sampling adult populations and prey dynamics. The two scientists chatted about Bonzek's pilot project and Latour's new grant and decided to team up. They agreed to use data analysis and field research to develop a new type of trawl survey on bigger adult fishes and predator-prey interactions. The study would include gathering information on aging and male-female ratios, which had never been recorded in the Bay, despite precedents around the world. They hired a research assistant, Jim Gartland, who was just finishing his master's degree at VIMS.

After the first three years showed excellent progress, in 2004 we extended the grant with an additional $195,000 for two more years. By that time, other Bay scientists, managers, and politicians were paying attention. VIMS was able to use the VEE funds to leverage funding from the NOAA Chesapeake Bay Office (NCBO) so that the program was fully supported through 2006. The program was firmly established at VIMS.

The Chesapeake Agreement's 2005 deadline for a plan was imminent, but now there was a basis for optimism that multi-species, ecosystems-based fisheries management was possible. A lot of nuts and bolts, literally, were needed to do this research: larger and more robust gear, such as bigger nets for trawl surveys. In order for VIMS's flagship vessel, the *R/V Bay Eagle,* to pull the nets quickly enough to catch large fish for the new survey, it had to be completely outfitted, including with newly fabricated plates and winch.

Years later, in 2018, VIMS took delivery of a brand new, state of the art research vessel, the *R/V Virginia.* The need and much of the design for this vessel was due to VIMS's expanded fisheries program. VIMS now had a top-quality sampling platform, and the role of this program in convincing the governor and the General Assembly rendered clear the need for this new $10 million vessel. The ecosystem and fisheries argument was a big selling point.

DID WE know how the VIMS ecosystem and fisheries research study was going to work out? Not according to Rob Latour, who is candid enough to say that they had little idea what they were doing at first. Along the way, they hit some lows in confidence. There was some skepticism because the VIMS researchers were relatively inexperienced, and no one had done this before. However, they proved themselves after embarking on this large-scale field operation, learning on the job and figuring out each new challenge as they went.

From the beginning, Dr. Latour encouraged the adoption of new technology such as the trawl net, which was equipped with hydroacoustic sensors to ensure that it was performing consistently each time it was deployed. This was not a common tool in 2001. This new technology allowed the researchers to be confident that the trawl net was reaching the bottom of the Bay and was opening wide enough to capture fish. The sensors provide real-time data on the trawl net's opening and bottom contact, so that researchers can calculate the area and volume sampled, information that is vital for estimating fish abundance. There was a lot of trial and error involved before the routines were set.

The first big trawl survey launched in March of 2002; since then, it has been conducted five times a year, from the mouth of the Bay all the way north to Annapolis, taking eighty to ninety samples along the center axis of the Bay, collecting data on all the different fish they catch. This produces a snapshot of the Bay fisheries each time. VIMS has the most intensive data set for the Bay, including its collections of small numbers of specimens to determine age and other biological characteristics. Since starting the survey, 500,000 animals have been processed. Nothing else from an academic institution on the East Coast, and certainly no research on an estuary, rivals this data set.

Despite being sanctioned by the political leadership of the Chesapeake Bay region, the team also had to contend with the state fisheries management agencies, who had the authority to regulate fish harvests and who would have to accept this new way of doing so.

MANY OF the Bay's most important species remain in jeopardy today. Oyster populations, for example, continue to suffer the effects of the overfishing that has occurred for decades. The oyster harvests in the 1970s look relatively abundant by comparison with the present, but they were lower than those recorded in the early twentieth century. Oyster populations have not yet returned to the 1970s level.

Every winter, Virginia and Maryland conduct a crab survey that is used to inform regulations for the upcoming fishing season. But there has been no formal, comprehensive stock assessment by the states since 2011, though VIMS would have valuable data to contribute to that evaluation if it ever happens. The assessment is far better than mere survey data, because it is a mathematical model that integrates the biology and population dynamics of the resource, thus providing estimates of fishing mortality and population sizes through time. So far, the states don't see funding stock assessments as a priority.

In 2022, the most recent annual winter dredge survey of the Bay's blue crab population, which was released in May 2023, found the least abundance of blue crabs since the first survey, conducted in 1990. Researchers reported an estimated 227 million crabs in the Bay. The previous low, which was in 2004, was 270 million crabs. The latest Winter Dredge Survey completed in early 2023 showed modest improvement to 323 million.[7]

In Virginia, crab season is open from mid-March through November. On June 28, 2022, the Virginia Marine Resources Commission unanimously agreed to change the restrictions already in place for commercial harvests from October 1 to the end of the season on November 30. Those limits will carry over for six weeks into the start of the 2023 season, four weeks longer than originally planned. With the Chesapeake Bay crab population at its lowest ebb in thirty years of monitoring, fishery managers warned that more restrictions affecting a broader swath of crab harvesters are sure to come.[8]

## Menhaden

In addition to the responsibility of regulating their respective state waters, Virginia and Maryland are beholden to the broader Atlantic coastal regime of management, because so many of the Bay's fish are "visitors."

The first species chosen by the Atlantic States Marine Fisheries Commission as the focus for developing a plan was menhaden, a small fish found in the Chesapeake Bay and along the entire Atlantic coast that is harvested commercially in large numbers for bait and for use in producing fishmeal and fish oil. Unlike all other fish in the Bay, menhaden have been managed for decades by Virginia's legislature rather than by the professionals in the Marine Resources Commission. This was the only fish in Virginia managed by politicians.

Implicit in developing a plan for menhaden management was the matter of putting an annual cap on the number of menhaden that can be sustainably harvested year after year, a limitation that the industry has historically resisted. Without knowing how the annual harvest affected the balance of menhaden left for future years, fishing proceeded unimpeded for years. The rationale for this was the maintenance of the two hundred or so jobs the menhaden industry supported. However, if the overfishing ever reached unsustainable, exhausting levels, it was argued, the jobs would disappear too.

The kind of economic and ecological analysis performed by Joanna Campbell in Staunton might have helped, but nothing like that was done. Instead, the regulation of menhaden was left to politics.

Dr. Latour and his team went about the tedious work of figuring out how to assess the size of the menhaden resources accurately and developing models for their sustainable management. Because of the prevalence of menhaden all along the Atlantic coast, the ASMFC, which has served the Atlantic coastal states for over seventy-five years, also became involved. The ASMFC regulates all recreational and commercial harvests of migratory fish from Maine to Florida and coordinates the conservation and management of twenty-seven nearshore fish species. Each coastal state is represented by three commissioners. Thanks in part to the work of Dr. Latour and his team, and also to advocacy by organizations such as the Chesapeake Bay Foundation and many others, in 2006 the ASMFC imposed strict caps on the harvesting of menhaden in the Chesapeake Bay and lowered them again in 2017.

Threatened with a moratorium on menhaden harvesting, in March 2020 Virginia enacted a law that assigned management of the menhaden fishery to the Virginia Marine Resources Commission, which manages all the other saltwater fisheries in Virginia. Just a few weeks later, the VMRC adopted a new regulation that required the 2020 menhaden harvest in the Chesapeake Bay be cut to almost half of the 2019 harvest level, and ASMFC withdrew the moratorium threat. That same year the ASMFC began for the first time to consider the management of menhaden on the basis of not only its ongoing sustainability but also its role as a source of food for other species.

In other words, it was not until almost twenty years after VIMS began its ecosystem-based analysis for managing fisheries, and fifteen years after a plan was supposed to be developed to manage Bay fisheries in a multi-species, ecosystem-based way, that fisheries managers finally began to use that approach. Now VIMS is among the acknowledged leaders of the most important new approach to managing the Bay's fisheries. Today, ecosystem-based, multi-species data management and planning is changing the harvest limits for menhaden along the Atlantic Coast and in the Chesapeake Bay.

By the end of 2022, thanks to the ecosystem-based technique developed by VIMS, the ASMFC concluded that menhaden were doing better than expected. It raised the menhaden Atlantic coast quota by 20 percent but took a more cautionary approach in the Bay, leaving the Bay catch at 51,000 metric tons.[9]

The adoption of limits by Virginia and by the ASMFC is just the beginning. More plans for other species are being developed. Together, science and data are the key to sustainable fisheries management and will set the course for other species to be managed for the long-term benefit of fisheries, fishers, and the environment. VIMS deserves great credit for this progress.

THE DEVELOPMENT of the ecosystem-based model for fisheries management by VIMS has had a national impact at the operational level of fish management. The VIMS team has evolved from a small group of recently graduated researchers to a full staff of thirteen people receiving weekly requests from managers and scientists, as well as from fellow modelers at work in New Jersey, Lake Erie, and Rhode Island looking for start-up advice. In addition to the program's continued scientific advancement, sixteen graduate students have been trained there to date and now teach at other institutions or serve on natural resources agency staffs, further multiplying the impact of the original VEE grants. The program now has a $2.1 million annual budget.

The Endowment's grants to VIMS were necessary to launch the research program, but, while generous, were not sufficient to sustain it. VIMS did, however, leverage federal money such as Wallop-Breaux Amendment funds from a tax on fishing equipment that is returned to states to use for sport fish restoration. It funds much of the VIMS ecosystem-based work now. As a result, the funding for this program has been predictable and stable since 2006.

With the adoption several years ago of milestones and a deadline for Bay cleanup results, the Chesapeake Bay finally has a deadline-driven set of commitments for the improvement of water quality.

The ecosystem-based, multi-species plan remains a work in progress. By and large, however, the advances in fisheries science and—more slowly—in fisheries management make it possible to determine how many fish there are in the sea and manage them sustainably.

FUNDING SCIENCE can be a way to support environmentally sound land use and conservation measures, but it is often necessary to "connect the dots" on behalf of grant-making organizations. When Chris Ludwig from the Virginia Department of Conservation and Recreation approached VEE about the Flora of Virginia Project, at first it seemed unrelated to our priorities. The idea was to update, document, and illustrate all the native and naturalized plants that grow from the Eastern Shore of Virginia to the mountains in the southwest.

We were, of course, intrigued to learn that this new flora (the name for a treatise on plants) would be the first one published since Thomas Jefferson roamed the woods of Virginia in 1762. It seemed like it was time for a new edition.

What convinced us was the Flora of Virginia Project's potential value to the Commonwealth's Natural Heritage Program, which we had helped to start many years earlier, as well as the strong likelihood that it would provide developers, land planners, and local decision-makers with information to help them conserve Virginia's natural resources, per their obligation to do so from Article XI. The Flora would also help teachers to meet the state's Standards of Learning for environmental knowledge. Even though collecting the data, drawing the illustrations, and developing an app would take a long time, VEE committed the first money to the project.

The Flora was published in 2012.[10] Although useful for all kinds of botanical field work, it is difficult to carry, weighing in at six pounds, so they developed an app.[11] The Flora of Virginia Mobile App "put the full contents of the print Flora into your pocket."[12]

## More Science in Service to Public Policy

Another example of science applied to public policy occurred in 1983 when VEE made a grant of $18,000 to the Virginia Institute of Marine Science for research to determine how chlorine residuals in sewage plant effluent affect oyster larvae. At that time chlorine was the principal agent used for disinfecting treated sewage wastewaters. Preliminary evidence suggested that chlorine residuals in receiving waters can adversely affect the development of oysters and other aquatic life. The studies done by VIMS provided data showing that the deleterious effects of chlorine residuals were contributing to the decline in oyster spatfall in the James River. Presented with these studies, the State Water Control Board adopted regulations limiting the use of chlorine in sewage treatment plants.

In 1989 VEE funded Dr. Clifford W. Randall's research at Virginia Tech's Department of Civil Engineering that developed improved design and operation of sewage treatment plants using biological nutrient removal (BNR) systems. BNR removes nitrogen and phosphorous from wastewater through the use of microorganisms and by adjusting environmental conditions in the treatment process. Virginia Tech was a leader in the development of this kind of solution.

# 21

# Climate Change

CLIMATE CHANGE is affecting the entire planet. That means it is also affecting individuals and the places that are special to them. The scale of climate change and its impacts, even on a single state like Virginia, can still be hard to wrap one's head around. Engineering education taught me to break large problems down into smaller pieces and solve them one by one.

Thinking about the problem of climate change and how to solve it, I remembered a story about the early-twentieth-century American actor and social commentator, Will Rogers. During the height of World War I, American ships crossing the Atlantic to bring supplies to our soldiers were being destroyed by torpedoes fired from the new German U-boats (undersea boats, or submarines). When Will Rogers was asked how he would solve this problem, he responded, "Well, it's simple. All you have to do is heat the Atlantic Ocean up to the boiling point and all these U-boats will pop to the surface and you can pick them off." And his interlocutor said, "Well, that is the stupidest thing I've ever heard. How could you ever do that?" Rogers replied, "Well sir, I just do policy. I don't do implementation." Some days the climate change problem feels that way.

It is admirable that in 2020 the Virginia General Assembly enacted the Clean Economy Act, thanks in part to the efforts of the Virginia Conservation Network and its partners. The successful implementation of it, however, is critical.

The Clean Economy Act sets out in law that Dominion, the largest electric utility in Virginia, must get to zero carbon by 2045 and APCO in the western part of Virginia has until 2050. No coal and no natural gas—nothing that emits carbon—will be allowed after those deadlines. Solar and wind power, along with whatever new technologies are developed, will make up the difference.

As Mary Rafferty of Virginia Conservation Network explains, carrying out these mandates involves numerous complexities:

> We got the clean economy act passed. We came up with the idea—the environmental community came up with the idea. We wrote the bill, and

for a change, Dominion had to follow our lead. We had brought the clean energy community and the environmental community together. And for the first time, they were working off our playbook, as opposed to them coming up with a big bill that we were just trying to make better.

But I think what makes the climate crisis complicated as well is that, in theory, we've tackled that source because, you know, we have the Clean Economy Act. But there's still a lot to figure out, like where are we going to site all of the facilities and power lines? There is also an energy efficiency provision in there intended to lower the overall amount of energy used.

And transportation is the largest source of carbon pollution. And that's why I bring up small things like pedestrian infrastructure or bike lanes or things that may not seem like they are big, massive climate action. But for every individual that we're getting out of a car, the carbon footprint decreases getting from point A to point B.[1]

It is not only the power companies and the transportation industry involved. Home builders, local governments, public transit, car manufacturers, and solar and wind energy companies are all involved. Mary Rafferty summarizes, "The big problem–when you start looking at how we're building our communities and how people are getting around and all of that–it's, all of a sudden, a bit more complicated."

In the early years of the twenty-first century, the concept of climate change was only just beginning to be addressed by those paying attention, who were mostly scientists. But the effects of climate change in coastal areas of Virginia were becoming too obvious to ignore. Rainfalls more often led to flooding in low-lying cities such as Portsmouth and Norfolk, erosion of beaches was creating legal headaches for once-permanent property lines, and authority to do anything about any of this was very limited under the laws in effect at the time. As is often the case, one person saw this need and proposed a solution.

It was Shana Jones who first envisioned the Virginia Coastal Policy Center. We met at the Miller Center at the University of Virginia during a conference on climate change and its effects in Virginia. Everyone was chatting during dinner, packed rather tightly together due to the arrangement of the table, when she floated the idea that we really needed a place for focusing on climate change in our state. Most people, she went on to explain, don't seem to understand the local ramifications of climate change in terms of the connections between property values and erosion, sea-level rise and flooding and zoning issues, and property shifting and all kinds of real, on-the-ground consequences.

What was needed, she said, was a joint effort involving both scientists and lawyers, a combination that could address the practical effects of climate change on shorelines, property values, and zoning implications, as well as science-based adaptive responses. Climate change, specifically sea-level rise, has been causing many environmental problems, and changing local shorelines, which in turn affects the everyday lives of lots of people.

Put differently, if water is coming up to your back porch in Mathews County, what does that mean for the worth of your land? The fact is that climate change has real-life implications for people living and working in Virginia's coastal areas. Denial is not a satisfactory strategy for addressing them. It is necessary to face up to how this new reality affects zoning laws, floodplain ordinances, and people's legal rights and responsibilities.

Ms. Jones approached the leadership of two schools at William and Mary: the School of Law and the School of Marine Science. Her idea that both science and legal knowledge had to be used together to deal with the complex nature of climate change's effects on Virginia's coastal region was a compelling one. She convinced the deans of both schools at W&M that this was a good idea and got everybody on board. By the time she came to VEE with a proposal, all the pieces were in place, and we said, "Yes, we can help with that"—and it was off and running.

William and Mary agreed to establish a new center called the Virginia Coastal Policy Center (VCPC), which Ms. Jones was named to lead. Once again it became clear why, rather than give money to institutions, VEE prefers to give money to individuals associated with institutions, allowing them the wider resources that institutions can provide to support their work. VCPC is a good example of that. Although based in the law school, drawing expertise from the School of Law and Marine Science, it focuses its work on tackling real world problems.

Roy Hoagland, a longtime environmental lawyer, succeeded Ms. Jones, and a few years ago Professor Elizabeth Andrews was named director of the Virginia Coastal Policy Center.

The Virginia Coastal Policy Center at William and Mary Law School has provided science-based legal and policy analysis of ecological issues affecting the state's coastal resources, providing education and advice to a host of Virginia's decision-makers, from government officials and legal scholars to nonprofit and business leaders. VCPC's approach to the effects of climate change is specific, local, and practical, involving lawyers, scientists, nonprofit representatives, local government leaders, and emergency management officials.

In February 2023, Professor Elizabeth Andrews announced that she would be departing the Center by June 30, 2023. On February 28, 2023, the dean of the William and Mary Law School announced that he had decided to close the Center, effective with her departure.[2]

# Epilogue

Those who never give up win in the end.

—Thomas Edison

AFTER WORKING for more than fifty years in environmental grant-making, here is what I have learned:

It takes a lot of people working together over a long period of time to make a difference in protecting and improving the environment.

Environmental laws can change often, sometimes by attempts to weaken them in response to new leadership and sometimes to add new ones and strengthen existing ones. In environmental legislation nothing is permanent.

Going green is a process, one that moves from taking air, water, and natural resources for granted in one's life to gradually realizing they comprise our life-support system.

Sometimes it takes a hurricane or other natural disaster to realize the importance of having safe water come out of your faucet. Sometimes what wakes you up is a poisoned water supply or a damaged system of water treatment plants, as people have endured in Flint, Michigan, and Jackson, Mississippi, respectively. As seen by the stories I have shared in this book, sometimes your own neighborhood is threatened and you realize you must act to protect your piece of the environment. Place by place, person by person, local action preserves the larger community.

The Governor's Council on the Environment was a strong proponent of the idea that government, as it goes about its business as the servant of the people, ought to pay attention to the public. Public hearings are now standard in regulatory proceedings, and Virginia citizens participate knowledgeably in the formulation and implementation of laws, policies, and rules to protect the environment.

In 1977 when the Endowment began, very few environmental groups existed. Now, Virginia has perhaps the nation's best nonprofit environmental infrastructure, and in the Virginia Conservation Network, its best statewide network. VEE provided initial, and often sustained, funding to a great many groups, such as the Institute for Environmental Negotiation, the James River Association, the Southern Environmental Law Center, the Elizabeth River Project, the Chesapeake Bay Foundation, The Nature Conservancy, the Environment Virginia annual conference, and VIRGINIAforever, among others.

VEE played an important role, but so did the citizens of Virginia who demanded political accountability for the protection of their environment from "pollution, impairment or destruction." That work continues today.

There is much to be grateful for. After fifty years of positive accomplishments, including new laws, policies, and programs, environmental quality in Virginia is in good condition and getting better. Municipal sources of drinking water are clean and reliable; water quality in rivers and streams is substantially improved; the air is visibly cleaner; poisonous chemical discharges into air and water have been restricted, documented annually, and monitored; a multistate "Save the Bay" program to protect the Chesapeake Bay is slowly being implemented; soil and chemical runoff fouling local streams has been reduced; and millions of acres of scenic and historic landscapes have been preserved. There has been a lot of progress, but the Clean Water Act's goal to eliminate pollution discharges remains unfulfilled, the 1987 Chesapeake Bay Agreement's call for zero discharge of toxic substances remains an elusive goal, and the realities of climate change demand attention.

Despite the negative rhetoric about environmental laws and regulations emanating from some politicians and business groups, the fact remains that the American economy is the strongest that the world has ever known and has been so ever since the end of World War II. Gross domestic product (GDP) is an accepted way to measure economic progress and power.[1] It represents the market value of the goods and services produced by labor and property located in the United States. At the end of World War II, value of the US GDP was 243.264 (each unit equals one billion dollars). At the end of the first quarter of 2022, its value stood at 24,386.734, which is the highest value among the countries of the world.

For polluters, who presume to have a right to dump poisons into the environment, to pose as victims of a "burdensome" regulatory apparatus reflects a sense of entitlement that is in direct conflict with the economic facts of life in the United States. It also contradicts the experience of the many corporations in Virginia who have received Environmental Excellence Awards for eliminating discharges. Decade after decade, the forces for loosening regulations continue to repeat their "costly, burdensome, and job-killing" excuses to avoid preventing pollution in the first place. Vigilance in monitoring and persistence in challenging these false prophets of doom and gloom will be necessary to maintain and improve environmental quality. This is because, in Virginia, thousands of permits are granted every year to those who apply to discharge all manner of poisons. Getting from permission to prevention takes persistence.

More people than ever are making their views known to the decision-makers who temporarily occupy positions of authority. This is what is driving environmental protection and improvement. These folks are "law and order" types when it comes to a clean environment.

Enforcing clean water laws is more complicated when it comes to more diverse sources of water pollution. Poison runoff from farms, fields, forests, and urban areas is more diffused than other sources of pollution, and therefore, cleanup was traditionally left largely to voluntary efforts subsidized with cash from taxpayers via state and local government programs. Farm subsidies to reduce poison runoff have recently been institutionalized. Results are still out on their effectiveness in reducing pollution of the Chesapeake Bay.

According to a 2022 estimate, at the current rate it could take another forty years for the Chesapeake Bay Program to achieve its 2025 objectives. Even then, money to match the rhetoric is going to have to be invested at higher amounts than previously, because of inflation.

With respect to pollution, the twenty-first century has the potential to become the Age of Prevention. Especially in the United States, known for its technological innovations, the likelihood of developing better ways of controlling pollution and eliminating discharges is high. As the Environmental Excellence Awards demonstrate, companies can use technology to eliminate discharges while also saving money, in effect refuting the argument that pollution controls cost too much.

Millions of people are insisting on this. Even small children are paying attention and already know what pollution and destruction of the environment looks like, and they don't like it. As they grow older, they will bring a new expectation of pollution prevention to the work of creating a society that values its natural resources, scenic vistas, and free-flowing rivers and streams. We know better now.

PHILANTHROPY HAS an important role to play in helping to move the country from permission to elimination to prevention of pollution. Private philanthropies in Virginia have been effective in helping people to make a difference. As neither government organizations nor private businesses, with their ability to fund programs that neither business nor government could justify, philanthropic grant-making foundations occupy a unique niche in this story. When they operate as seed capital for social change, they can leverage relatively small investments into major public policy changes. There are foundations in every state that can help in this way.

Philanthropy, when it is done with a focus on results, is like a venture capital fund for catalyzing social change. Organized philanthropy is "patient money," meaning that it can afford to engage in issues for the long run. Foundations remain one of the great potential sources of progress in our fragile democracy.

Grant applicants are understandably nervous when asking a foundation for money. The Endowment tried to see these requests as an opportunity to learn from someone "on the ground." Betty Toler, VEE's assistant director for our first twenty-two years, was particularly good at greeting first-time visitors and making them feel comfortable and VEE more approachable. And if some people needed help to walk their way through the grant process, we were happy to help, because that's what our job was, to spend the money well. I have never met anyone better than Betty at that; she is a great people person.

Our philosophy: get to know the people asking for grants and get to know them better once you make a grant. Some of our best grants happened that way. We didn't give money to strangers.

The Endowment helped to establish donor networks in which foundations can collaborate on funding. As discussed in chapter 10, VEE was one of several Chesapeake Bay–area foundations that founded the Chesapeake Bay Funders Network of foundations who support environmental research, education, and advocacy to restore the Bay. They are based in Maryland, Pennsylvania, Virginia, and the District of Columbia, and because of their collaborative funding to many Bay-area groups, they have been influential in driving the Bay Program toward achieving its 2025 water quality objectives. A special tip of the cap to Verna Harrison, former Assistant Secretary of Natural Resources in Maryland, a principal author of the 1987 Bay Agreement's "Living Resources" chapter, and the guiding and driving force behind the Bay funders network while president of the Campbell Foundation.

One illustration of the power of such collaboration and the leverage possible by focused grant-making is the pilot project jointly funded by the Campbell Foundation and the Virginia Environmental Endowment to fence cattle out of streams. This began as a demonstration project to show that it could be economically feasible for farmers to erect fences along streams, thus keeping soil, fertilizers, and animal wastes on land rather than polluting the streams. In Virginia this has developed into the multimillion-dollar annual subsidy known as the Agriculture Cost-Share Program, which supports farmers to accomplish these goals. Planting trees and plants as buffers is also now included in the program.[2]

TODAY, VEE continues to make a difference, and the ripples from Judge Merhige's initial decision continue to extend far beyond his lifetime. The Endowment currently operates three programs: the Virginia Program, the James River Water Quality Improvement Program, and the Community Conservation Program.[3] Current grant-making priorities in the Virginia Program are focused on improvement of local rivers and protection of water quality, restoration of the Chesapeake Bay, innovative land conservation and sustainable land use practices, environmental literacy and public awareness, and climate adaptation. The James River Water Quality Program came about in April 2018, when VEE announced the creation of a new grant program totaling more than $15 million, designed to accelerate and advance significant water quality improvements throughout the James River watershed. The James River has made remarkable improvement over the past forty years, but there remains considerable need for reductions in nutrient, sediment, and chemical pollution if we are to fully restore the river. VEE's "Community Conservation Program" provides funds for conservation initiatives located within the counties of Craig, Franklin, Giles, Montgomery, Pittsylvania, and Roanoke, as well as in the cities of Salem and Roanoke. The Program focuses on water quality protection, restoration, and improvement; land conservation support; and environmental literacy and awareness.

In 2017 the Endowment launched a year of special activities celebrating VEE's fortieth anniversary, culminating at the Virginia Historical Society on October 5 with the recognition of VEE's twenty-two Partners in Excellence.[4] Over its forty-five-year history, VEE has partnered with nearly five hundred nonprofit organizations, universities, government agencies, schools, and communities. Of these, twenty-two were selected as Partners in Excellence honorees and awarded $1,000 in recognition of their contributions to Virginia's environment. In addition, VEE made a $10,000 grant to a documentary being developed by a nonprofit film organization on the late US District Court Judge Robert R. Merhige Jr. as a special recognition for his role in establishing the Endowment.[5]

Overall, the importance of the VEE story is how it was able to help hundreds of groups and thousands of people to make a difference in the quality of their environment. The people are the ones who most deserve the credit for the progress of recent decades.

In Virginia the successful advocacy of better policies, laws, and practices by the Virginia Conservation Network, its members, and its partners has produced new laws and greater awareness of the need to fund their implementation, turning the ship of state's direction. By comparison with foundations in general, the Virginia Environmental Endowment's pattern of sticking with people and grantee organizations for a long time is unusual. This notable pattern is visible with several signature grants, like those awarded to the Southern Environmental Law Center, the Institute for Environmental Negotiation, the Chesapeake Bay Foundation, The Nature Conservancy, the annual Environment Virginia conference, and several "friends of the river" groups. The first time I remember that happening was during a board meeting in our second or third year. A grant request to renew our support was on the agenda, and one board member asked, "Are they doing a good job?" I said that they were, and a motion was immediately made to renew the grant. I doubt that adopting the "three years and out" funding rule that some foundations follow would have produced the results that these organizations have achieved.

The Endowment has leveraged its grants by requiring matching funds. We have focused on first-dollar funding to help launch innovative ideas and new groups. We want to see you in person, discuss your ideas, and give you the opportunity to make a successful grant pitch. We have welcomed everyone who shares the goal of improving the environment to be a part of this movement. In the words of longtime VEE board chair, Dixon Butler, "We are passionate advocates for the middle of the road."

Our first poll in 1995, and subsequent ones as well, showed that environmental justice was a hidden issue. One had to parse the poll details to see it, but it was there. One of the ways it came out clearly was that the strongest demographic for protecting the environment in Virginia at that time was Black people. Why? Perhaps because so many African Americans live with the consequences of pollution daily. Many felt strongly about the environment because they had experienced the negative effects of pollution and were struggling to stay alive or avoid being poisoned. When Lois Gibbs organized her neighbors to fight the poisoning of their lives in Love

Canal, New York, environmental contamination was also a justice issue. After she relocated to Northern Virginia, the Endowment helped her in 1985 to start the Citizen's Clearinghouse for Hazardous Wastes and again in 1992 for activities in rural Floyd County, Virginia. We helped her both in Virginia and through our Kanawha and Ohio Rivers Water Quality Program in West Virginia.

The first two surveys of public opinion that we commissioned both showed overwhelming support for environmental quality, and it was consistent across the state and with every demographic. We also observed, however, that when it came time to vote, this strong support did not always translate into electing the most environmentally friendly statewide candidate.

We took another poll to try to find out why. The result showed that the closer the contest is to the local level of government—town, city, and county—the more important the environment became as an issue. However, voters in southwest Virginia's coal mining region tend not to worry much at all about traffic congestion in the Washington, DC, suburbs located in Virginia's northeast corner. Sea-level rise in Norfolk is not much of a priority for people trying to preserve farmland in central Virginia. Scaling up in electoral district size to congressional districts and statewide races, the polls showed that it is because local issues are not statewide issues that the environment becomes less of a factor compared with other issues. This is why it is crucial to think globally but act locally. National governments should continue to gather for debate how to deal with the effects of climate change, but specific steps need to be taken toward eliminating discharges and emissions right near our own homes. Eliminating those sources of pollution will go a long way to eliminating them worldwide. As Faye Cooper of the Valley Conservation Council put it: "I believe climate change is the biggest environmental threat to ever face the earth. There is a role for every environmental group, including land trusts, to engage in finding solutions such as permanent protection of forests and grasslands for carbon sequestration in a deliberate and targeted manner."

In 1970, there was no effective infrastructure of environmental groups in Virginia. Over the past fifty-three years, I have witnessed, helped, and encouraged many people, both professionals and volunteers, to figure out what needed doing to improve the environment and marshal the resources to do it. One of the Virginia Environmental Endowment's legacies is the many groups we helped to launch. It was VEE's funding and support for the nonprofit conservation organizations that allowed them to persist in their advocacy. Collectively, since 1970 they have made tremendous progress one step at a time, sometimes two steps forward and one step back, depending on how the political winds were blowing. Overall, though, the environment is in better condition, stronger laws are in place, and most importantly, today there is a well-established group of nonprofit public advocacy groups to hold polluters and regulators accountable. That is a major difference.

Environmental protection, whether it's done by SELC or the Valley Conservation Council or the Elizabeth River Project, is premised on conserving a sense of place. All the good that has resulted came about because somebody cared about where they lived and worked and said, "No, I'm not going to let anyone ruin this place." That is what drove them in the first place: the potential for losing a place that was special to them.

It is the people and their stories that make protecting the environment possible: Rick Middleton, Deborah Murray, David Carr, Geoff Gisler, Trip Pollard, and Kay Slaughter at SELC; Faye Cooper at the Valley Conservation Council; Will Baker, Joe Maroon, Ann Jennings, Roy Hoagland, and Peggy Sanner at Chesapeake Bay Foundation; Patti Jackson and Bill Street of the James River Association; Michael Lipford and Nikki Rovner at The Nature Conservancy; Ann Regn, Suzie Gilley, Lisa Deaton, and Tamra Willis, who promoted environmental education throughout Virginia; Wetlands Watch's Skip Stiles, who teaches people about climate change and rising sea levels' effects; Marjorie Mayfield Jackson and Robin Dunbar of the Elizabeth River Project; Ann Swanson of the Chesapeake Bay Commission; the women of the Garden Club of Virginia, who have done so much for so long to advocate for the environment; and so many more who have devoted recent decades to improving Virginia's environment.

The next generation of leaders continue the work of environmental conservation: Mary Rafferty of the Virginia Conservation Network; Hilary Harp Falk, new President of the Chesapeake Bay Foundation; Adam Schellhammer at the Valley Conservation Council; Kate Wofford of the Alliance for the Shenandoah Valley; Wendy Austin and Heather Barrar at the Friends of Lower Appomattox River; DJ Gerken at the Southern Environmental Law Center; and Angie Rosser of the West Virginia Rivers Coalition, to name just a few.

Leading most people's lists of the issues they choose to engage in are climate change and clean energy reforms. Healthy rivers, saving the

Chesapeake Bay, land conservation, and green communities are priorities that will continue to define the challenges of the current century. Nonpoint sources of pollution from both urban and rural areas remain a problem requiring state money. Employment in solar and wind energy has grown rapidly and is outpacing traditional carbon-based industries. Prevention of pollution might finally overtake permission in this century.

Finally, I want to acknowledge how year after year, thousands of people throughout Virginia, the Chesapeake Bay region, West Virginia, and Kentucky volunteer their time, talent, and treasure to help. It all adds up. You may not be able to influence other countries, or even other states, but you can do something wherever you live. That has worked well here. That idea is powerful.

Lots of people have worked together over the years to get us to where we are today, and if you had asked many of them when they were children if saving the environment might be their life's work, I doubt that many of them would have said yes. It never occurred to me, for example. Yet, there are many people we can thank for volunteering to protect their special places. The ranks of people who work full-time in science, engineering, public policy, and natural resources management have increased dramatically too. Colleges and universities now have majors in environmental topics that did not exist fifty years ago. There are more parks and nature centers than ever. Government and private organizations have dramatically increased their employment of environmental specialists. There is a large group of public advocacy groups and other nonprofits in place today, and they are here to stay. People do not want their environment polluted. Period.

For those of you who might be new to all this and would like to help, there are plenty of opportunities and there is much work remaining to be done. I hope you will build on the progress to date and do even better going forward. We are much better prepared to face the environmental challenges of the next fifty years. You do not have to save the entire world to make a difference. Just take care of your part of it.

AFTER A half a century, we are still permitting discharges under a law that is intended to eliminate them. When the United States Senate was discussing the Clean Water Act, I was there. I saw what they said, and it was clear that they meant: stop, no more discharges, eliminate them. To paraphrase the late English writer and philosopher G. K. Chesterton's thoughts about Christianity, the problem with the Clean Water Act is not that it has been tried and found wanting, but that it has been found difficult and left untried.[6] After fifty years, the goal of the Clean Water Act to eliminate discharges is still the law and remains unfulfilled.

As far as the Clean Water Act and the National Pollution Discharge Elimination System (NPDES) goes, what developed instead was a national pollution discharge *permission* system. In practice, the mandate to eliminate discharges has been eliminated. That suggests that there are hundreds of thousands of dischargers, who are violating at least the spirit of the law, if not the law in fact. How about the government doing what the Clean Water Act requires? How about: "No more discharges will be permitted, pursuant to the Clean Water Act." It is past time to follow the law. The many companies who won Environmental Excellence Awards show that this can be done.

And to those whom I have met who say, "That's not realistic, the system is embedded now," and express similar thoughts, I respond that if President Jack Kennedy had thought like that, we would never have landed a man on the moon. If General Dwight Eisenhower were "realistic," he never would have proceeded with the D-Day landings. If Judge Merhige had been following the prescribed rules, he never would have created the Virginia Environmental Endowment.

There is something to be said for the concept of right action, meaning in this case to eliminate poisonous discharges rather than continuing to permit them. Kevin W. Irwin, in his book about Pope Francis's 2015 encyclical *Laudato si': On Care for Our Common Home*, points to the significance of the subtitle, noting that the encyclical is addressed to all the people of the world and "invites us to action to *care* for it . . . to 'act' as responsible citizens . . . in the public square and in our personal lives, in the corridors of political power and influence. . . . [T]he pope again and again refers to the import of the *common good* . . . over an *individualism* that . . . infects our contemporary culture. . . . *common home*, a home for plants, animals, and humans."[7] The Clean Water Act, if carried out as originally intended, embodies that hope for humankind.

MOST OF the people I have had the honor to work with all these years did not consider the difficulties that might arise; they just went ahead and did whatever they could to improve their special place. Whether it was Virginia's inaugural public hearings on the environment, developing an action agenda based on the Commonwealth's first report on the status of Virginia's environment, launching an environmental mediation center to resolve complex disputes, or funding public interest lawyers to advocate before regulatory boards, it was all aimed at improving the quality of the environment. Furthermore, the things that the people demanded of their government were practical, such as clean water and air, new state parks, and eliminating toxic discharges into air. People led the efforts to get things done.

And it was the Endowment's polling, which consistently showed strong support for a clean environment, that helped elected representatives to support environmental measures.

WHEN "WE the people" demand action to improve the environment—or any other social good—things start to happen. There has been great progress over the past fifty years. That does not mean we have finished; there is much left to do, and as Governor Holton told me a long time ago, change is incremental. When VEE began its work in 1977, neither I nor the board of directors foresaw the tremendous accomplishments that hundreds of groups and thousands of people would achieve with a little help from the Endowment. Once again, we are reminded that little of lasting value is accomplished alone. There is one evergreen fact of life in environmentalism: it takes a lot of people working together over a long period of time to make improvements.

The Endowment stood on the shoulders of wonderful leaders in environmental conservation who came before us.

Now, it is your turn. Go and do likewise.

# ACKNOWLEDGMENTS

I am deeply grateful to three writers who generously taught me so much about the craft of writing. Greg Smith helped me conceptualize the outline of the book; Constance Costas helped me move the narrative from telling to showing the action; and Whitney Roberts Hill helped to make the uniqueness of the story more apparent and personal. I am still learning their lessons. Any shortfall in these respects is completely on me. Melissa Scott Sinclair helpfully reviewed an early chapter of this book and made several constructive suggestions. Dean King spent time teaching me how to use stories to convey ideas. Big thanks to the James River Writers, the friendliest and most helpful group of writers anywhere. In the early stages of research and writing, the Virginia Historical Society, now renamed the Virginia Museum of History and Culture, allowed me frequent use of its E. Claiborne Robins Jr. Research Library and its Norfolk Southern Rare Book Room. My thanks to Paul Levengood and Canan Boomer for their many kindnesses to me. I wish to thank those people who took the time to be interviewed for this book: Sharon Adams, Will Baker, Joanna Campbell, Faye Cooper, Tom FitzGerald, Jim Gartland, Geoff Gisler, Marjorie Mayfield Jackson, Rob Latour, Tom Lewis, Rick Middleton, Deborah Murray, Lundy Pentz, Trip Pollard, and Mary Rafferty. In addition to the people whose stories I have quoted, many other people encouraged me along the way: Heath Hardage Lee, Larry Sabato, Dan Ludwin, Jeremiah Winters, Cale Jaffe, Birch Douglass, Ed Isaacs, Ann Regn, Kristin Beauregard, MM Finck, Karen Chase, Wayne Dementi, Mary Ellen Stumpf, William H. Carr, A. E. Dick Howard, W. R. Provo, April Eberhardt, Elizabeth Plaushin, Anna Lou Schaberg, Heather Mazeika, Henry Berman, Bud Minor, Kim Hulcher, John Farmer, Anna Lawson, Ayla Palermo, Angela Dugan, Nella Barkley, Joseph Maroon, Bill Mims, Tony Proscio, Kimberly Gluck, Terri Cofer Beirne, and the great team at the University of Virginia Press. As always, my family were constant supporters, early readers, and cheerleaders for this effort. To all these people, thank you!

# AUTHOR'S NOTE

I was an engineer whose interests turned to environmental quality. I have spent most of my career in the government and in philanthropy. This perspective and experience have allowed me to see how important the role of nonprofit environmental conservation groups has been to the improvement of environmental quality, not just in Virginia but throughout the country and the world. During a long career in environmental conservation, I have had numerous conversations about this work with younger people and have encouraged them to get involved or remain a part of it. I have delighted in telling them stories that answer their questions. Out of kindness, I suppose, many have asked, "How did you go from nuclear engineer to defender of the environment?"

As discussed in chapter 14, I still remember being annoyed as a child watching raw sewage being dumped into the local river and feeling that this wasn't right. Although my formal education and military experience had little to do with environmental topics, I maintained this curiosity about why dumping sewage and poisons in the water was even allowed.

At age twenty-six, as my discharge date from the Air Force was approaching, I had no idea what I wanted to do for a job or how to find one. With a baby daughter and another child on the way, I felt the urgency of getting on with it. I had read some career advice columns in the *Air Force Times* about how active-duty military could transition to civilian work that would use the considerable skills most of us had developed during our assignments. The columnist's suggestions made good sense to me.

The author was a Virginia-based career counselor named John Crystal, and I became one of his clients. John was one of the most brilliant and unorthodox people I have ever met. He helped people leaving the Armed Services find meaningful careers in civilian life, not by serving as an employment agency but rather by sharing a systematic process for capitalizing on the many executive skills they had acquired in service to our country. Instead of the "meat markets" that Fortune 500 companies invited junior officers to attend, where I never met anyone who could hire me—only screen people out—John took a strategic approach to career planning (not just a job search) that involved identifying each client's unique combination of skills, experience, and interests and matching those to organizations that shared those interests and needed such people. It was a lot more interesting and complicated than filling out job applications and sending out dozens of resumes. Instead of using resumes, John encouraged clients to make a proposal to the exact person who can hire you to be a part of the organizations you have identified as meeting your own criteria for employment.

John's approach led me to a months-long period of reading, researching, and thinking about the fields I was most interested in, which then led me to the environment as the field most important to me. I also decided I wanted to live in Virginia and moved here.

There was more to my coming to work for Governor Holton than the courtesy letter that precipitated that result. I had spoken in person about environmental problems with dozens of private and governmental experts and had become quite knowledgeable about environmental problems. With many senior officers flying missions in Vietnam, the Air Force had, of necessity, given many of us junior officers plenty of high-level management experience developing multimillion-dollar proposals, negotiating contracts with defense industry executives, managing the grants, reporting to the generals in the Pentagon and at the Defense Advanced Research Project Agency, and in short, carrying out assignments our contemporaries in civilian jobs would not get for years.

When I moved to Virginia in 1970, I visited John Crystal. He offered me a spare desk at his office to have a business-like base for my search. It was in a small strip shopping center in McLean at the original Tyson's Corner, a tiny spit of land that is only a memory now. One day, after I had just hung up the telephone with a prospective employer, John walked into the room. He smiled and reassured me that I was on the right course and that I would find what I was looking for. Then, as he waved his empty cigarette holder around, he said, "We have a new governor here in Virginia. Why don't you write a letter to him, let them know you're here?" I asked him, "Why would I do that?" He replied with a knowing grin, "Why not? You never know where opportunity lies."

# APPENDIX

## Further Information

On Thursday, October 7, 2010, I spoke about the Virginia Environmental Endowment in the Virginia Historical Society's Banner Lecture series. This lecture focused on the origins, mission, and accomplishments of VEE. The lecture examined the effects of the Endowment's grants on Virginia's environment and the people who have helped to make those results possible. This lecture addressed each of these aspects of its work and the strategic approach to grant-making that has made VEE a leader in environmental grant-making. It is available online at https://virginiahistory.org/learn/historical-media/virginia-environmental-endowment-leadership-leverage-and-legacy.

The Virginia Historical Society has changed its name to the Virginia Museum of History and Culture and maintains an extensive archive of materials donated by Virginia Environmental Endowment and others in its Robert R. Merhige Jr. Archives. For more information, see https://virginiahistory.org/research/research-resources/finding-aids/virginia-environmental-endowment-part-robert-r-merhige-jr-environmental-history-archive.

## Board Members Appointed by Judge Merhige

Ross P. Bullard, William B. Cummings, Cathleen H. Douglas, Frances A. Lewis, Sydney Lewis, Henry W. MacKenzie Jr., George L. Yowell, Thomas K. Wolfe Jr., Dixon M. Butler, Virginia R. Holton, Paul U. Elbling, Jeannie P. Baliles, Byron L. Yost, Patricia Kluge, Alson H. Smith Jr., Robert B. Smith Jr., Robert M. Freeman, Linwood Holton, Nina Randolph, and Robin D. Baliles.

Additional directors I served with during my tenure include Anna L. Lawson and Landon Hilliard.

VEE's wonderful staff during my tenure included Betty L. Toler, assistant director during our first twenty-two years, administrative assistants Jayne M. Mitchell, Joan Bragg Day, Barbara F. Clatterbuck, and for my final fifteen years, Jean Wildbore. They made the Endowment's office efficient and friendly to all who approached us.

# NOTES

## Introduction

1. Virginia Environmental Endowment, first annual report, 1978.

## 1. A Poisoned River

1. For more thorough reporting on the Kepone story, see Richard Foster, "Kepone: The Flour Factory," *Richmond Magazine*, July 8, 2005, https://richmondmagazine.com/news/kepone-disaster-pesticide/. And for a thorough examination of the Kepone disaster and its aftereffects, see Gregory S. Wilson's *Poison Powder: The Kepone Disaster in Virginia and Its Legacy* (Athens: University of Georgia Press, 2023).
2. To learn more , see https://law.lis.virginia.gov/vacodefull/title32.1/chapter6/article9/.
3. See United States v. Allied Chemical Corp., 420 F. Supp. 122 (E.D. Va. 1976), https://law.justia.com/cases/federal/district-courts/FSupp/420/122/1738722/. For an excellent explanation of how the federal proceedings against Allied and Life Sciences developed against the background of new chemicals coming into use and the rising tide of opposition to the danger to people and the environment, see Robert R. Merhige Jr., Manning Gasch Jr., William B. Cummings, Robert H. Sand, Robert B. Smith III, and W. Wade Berryhill, "Allied Chemical, the Kepone Incident, and the Settlements: Twenty Years Later," *University of Richmond Law Review* 29, no. 3 (1995): 495–520, https://scholarship.richmond.edu/cgi/viewcontent.cgi?article=2134&context=lawreview.

## 2. The Judge's Trust

1. See https://en.wikipedia.org/wiki/Robert_R._Merhige_Jr.
2. Merhige et al., "Allied Chemical," 493.
3. Bill McAllister, "Allied's Fine Cut to $5 Million for Kepone Pollution," *Washington Post*, February 2, 1977.
4. Merhige et al., "Allied Chemical."

## 3. An Independent Board

1. Merhige et al., "Allied Chemical," 509.

2. For more about the EPA, see https://ega.org/about.
3. In 1985, the Lewises gave most of their private collection to the Virginia Museum of Fine Arts (VMFA). VMFA's Sydney and Frances Lewis Galleries now represent one of the top postwar collections in a comprehensive museum in the United States.

## 5. ARTICLE XI

1. Howard, A. E. Dick. *Commentaries on the Constitution of Virginia,* vol. 2 (Charlottesville: University Press of Virginia,1974), 1154.
2. Howard, *Commentaries,* 1145–46.
3. Robb v. Shockoe Slip Foundation, 324 S.E.2d 674 (1985).
4. For an excellent analysis of Article XI and its unfulfilled mandate, see Tyler Demetriou, "Reinvigorating the Virginia Constitution's Environmental Provision," *Virginia Environmental Law Journal,* no. 40 (2022): 66–101, https://papers.ssrn.com/sol3/papers.cfm?abstract_id=4077369.
5. Montana Environmental Information Center v. Department of Environmental Quality. See https://casetext.com/case/meic-v-dep-of-env-quality.
6. Katy Spence, "Montanans' Right to a Clean and Healthful Environment," Montana Environmental Information Center (blog), April 29, 20120, https://meic.org/montanas-right-to-a-clean-healthful-environment/.
7. Professor Howard, the author and director of the commission that drafted Article XI, has spoken many times about the new constitution and this important mandate. See Mike Fox, "50 Years after Leading Constitution's Revision, Professor Reflects on Changes: A. E. Dick Howard '61 Helped Draft, Ratify Document," University of Virginia School of Law, November 11, 2020, https://www.law.virginia.edu/news/202011/50-years-after-leading-constitutions-revision-professor-reflects-changes.

## 6. PERMISSION TO POLLUTE

1. Federal Water Pollution Control Act of 1972, Section 101.
2. 33 U.S.C. § 1342.

## 7. THE CLEAN WATER ACT IN VIRGINIA

1. Patrick Wilson, "Va. Senate Democrats Pass GOP Bill to Remove Power of Citizen Environmental Boards," *Richmond Times-Dispatch,* February 15, 2022, https://richmond.com/news/state-and-regional/va-senate-democrats-pass-gop-bill-to-remove-power-of-citizen-environmental-boards/article_3a0a60df-fea9-584a-984e-88de0a8be1ef.html.

2. Sandy Hausman, "State Lawmakers Target Citizen Air and Water Boards," WVTF, January 28, 2022, https://www.wvtf.org/news/2022-01-28/state -lawmakers-attack-citizen-air-and-water-boards.

3. "Virginia Environmental Excellence Program," Virginia Department of Environmental Quality, accessed March 15, 2023, https://www.deq .virginia.gov/get-involved/pollution-prevention/virginia-environmental -excellence-program.

## 8. The Clean Water Act

1. Paul Thompson, Robert Adler, and Jessica Landman, *Poison Runoff: A Guide to State and Local Control of Nonpoint Source Water Pollution* (New York: Natural Resources Defense Council, 1989).

## 9. Environmental Policy in Virginia

1. Governor's Council on the Environment, "Virginia's Environment," 1971.

2. "From the Archives: James River Crusader Newton Ancarrow," *Richmond Times-Dispatch*, August 13, 2017, https://.com/discover-richmond/from -the-archives-james-river-crusader-newton-ancarrow/article_1fdf284b -7fee-5e5f-a918-bab0725629b4.html.

3. See the opening section of the National Enviornmental Policy Act website at https://ceq.doe.gov/.

## 10. The Updated Mission of the Endowment

1. Tom FitzGerald, email message to author, October 10, 2022.

2. For more information about the Kentucky Resources Council, see https:// www.kyrc.org/.

3. To learn more about the West Virginia Rivers Coalition, see https:// wvrivers.org/.

4. See the Chesapeake Bay Foundation's website at https://www.cbf.org/.

5. More on Friends of Dragon Run is available at https://www.dragonrun.org/.

6. For more about Virginia Humanities, see https://virginiahumanities.org/.

## 11. The Public Interest

1. Rick Middleton, interview with author.

2. "Newport News Scraps the King William Reservoir," Southern Environmental Law Center, September 23, 2009, https://www.southernenvironment .org/news/newport-news-scraps-the-king-william-reservoir/.

3. See https://www.oyez.org/cases/2006/05-848.

4. Phone interview with Deborah Murray. Alliance to Save the Mattaponi v. US Army Corps of Engineers, 606 F.Supp.2d 121. US District Court, District of Columbia.

5. Alliance to Save the Mattaponi v. US Army Corps of Engineers.

6. To read the full statement, see https://www.southernenvironment.org/staff /trip-pollard/.

7. See https://www.southernenvironment.org/wp-content/uploads/2023/04 /Beyond-Asphalt-Creating-a-Better-Transportation-Future-for-Virginia -SELC-report-1999.pdf.

8. For more on SELC, see https://www.southernenvironment.org/.

## 12. The Chesapeake Bay

1. To learn more on the Chesapeake Bay Foundation, see www.cbf.org.

2. Governor's Commission on Virginia's Future, *Toward a New Dominion: Choices for Virginians,* December 1984.

3. Chapter 20, on fisheries management, describes how this disconnect was overcome for the long-term management of menhaden.

4. The history of all the agreements adopted to restore the Bay is available online: https://www.chesapeakebay.net/who/bay-program-history.

## 13. Clean Water and the Growth of Environmental Advocacy

1. More information on the James River Association is available at https:// thejamesriver.org/.

2. *Norfolk Virginian-Pilot,* February 9, 2001; *Richmond Times-Dispatch,* February 9, 2001; *Roanoke Times,* February 9, 2001.

3. "Streams Cleanup Will Be Costly," *Richmond Times-Dispatch,* February 8, 2001.

4. More on Friends of the Rivers of Virginia can learned at https://www.forva .org/.

5. Marjorie Jackson, interview with author.

6. Information about the Elizabeth River Project is available at https:// elizabethriver.org/. Margaret Mead, *The World Ahead: An Anthropologist Anticipates the Future* (New York: Berghahn Books, 2004).

7. To learn more about Lynnhaven River NOW, visit https://www.lynnhaven rivernow.org/.

8. See https://www.cleanmyrivers.com/.

9. For more on the Dan River Basin Association, see https://www.danriver .org/.

## 14. Land Conservation and the Growth of Environmental Advocacy

1. Governor's Council on the Environment, *The State of Virginia's Environment, An Analysis and Recommendation,* December 1971, 45. (Report in author's collection.)
2. See, for example, https://www.williamsburgva.gov/173/Dillon-Rule.
3. The Land Preservation Tax Credit program is enabled by the Virginia Land Conservation Incentives Act. Through this program, Virginia allows an income tax credit for 40 percent of the value of donated land or conservation easements. In 2020, taxpayers may use up to $20,000 per year. Tax credits may be carried forward for up to ten years after the year of donation. Unused credits may be sold, allowing individuals with little or no Virginia income tax burden to take advantage of this benefit.
4. These words of Larry Selzer's are quoted from TCF's website: https://www .conservationfund.org/.
5. Faye Cooper, interview with author.
6. More can be learned about the Valley Conservation Council by visiting https://valleyconservation.org/.
7. For more about VaULT, see https://vaunitedlandtrusts.org/.
8. See https://lis.virginia.gov/cgi-bin/legp604.exe?991+ful+CHAP0900&991 +ful+CHAP0900.
9. For more on the Virginia Land Conservation Foundation, visit https:// www.dcr.virginia.gov/land-conservation/vlcf.
10. For more information, visit https://www.virginialandcan.org/local-resources /Virginia-Land-Conservation-Foundation/37268.
11. To learn about the Capital Region Land Conservancy, see https://capital regionland.org/.

## 15. The Virginia Conservation Network

1. Mary Rafferty, interview with author.
2. See https://vcnva.org/ for more about the Virginia Conservation Network.

## 17. Attitudes of Virginians about the Environment

1. Edwin G. Dolan, *TANSTAAFL: the Economic Strategy for Environmental Crisis* (Holt, Rinehart and Winston, 1971).
2. Edwin G. Dolan, *TANSTAAFL: The Economic Strategy for Environmental Crisis* (New York: Holt, Rinehart and Winston, 1971), vi.
3. Dick Morris, *Report on Survey of Opinions on Environmental Issues in Virginia* (Virginia Environmental Endowment, 1995).

4. See for example, the *Richmond Times-Dispatch,* June 21, 1995, 1.
5. *Virginian-Pilot,* June 23, 1995, A14; *Newport News Daily Press,* June 22, 1995.
6. *Fredericksburg Free-Lance Star,* October 30, 1995.
7. *New York Times,* July 5, 1995, A11; *Newsweek,* July 15, 1996, 26.

## 18. FOLLOW THE MONEY

1. Department of Environmental Quality; Department of Conservation and Recreation; Department of Game and Inland Fisheries; Virginia Marine Resources Commission; Department of Historic Resources; Museum of Natural History; Chesapeake Bay Local Assistance Office; and the Chippokes Plantation Farm Foundation.
2. "Impaired water" means water that is not meeting one or more state water quality standards, as required by the Clean Water Act; water with fish or shellfish harvesting prohibition by the Virginia Department of Health; and/or water where biological monitoring indicates moderate to severe impairment and is listed by stream segment on Virginia's 303(d) Total Maximum Daily Load Priority List.

## 19. ENVIRONMENTAL EDUCATION

1. I am grateful to Joanna Campbell and Dr. Pentz for helping me tell this story.
2. Lundy Pentz, interview with author.
3. For further details on the Virginia Junior Academy of Science, see https://vjas.org/index.html.
4. The Virginia Association for Environmental Education can be found at https://vaee.wildapricot.org/.

## 20. SCIENCE MATTERS

1. See, for example, https://www.chesapeakebay.net/.
2. The recent report of the Chesapeake Bay Stock Assessment Committee provides more information. See https://www.chesapeakebay.net/documents/CBSAC_2018_Crab_Advisory_Report_Final.pdf.
3. To read the entire document, see https://d18lev1ok5leia.cloudfront.net/chesapeakebay/documents/cbp_12081.pdf.
4. For those readers with a scientific bent, Dr. Edward Houde has published an excellent introduction to the ecosystem-based management approach: *Managing the Chesapeake Bay Fisheries: A Work in Progress* (College Park: Maryland Sea Grant College, 2011).
5. See https://www.vims.edu/ for more about the Virginia Institute of Marine Science.

6. Rob Latour, interview with author.
7. "Va.'s Crab Numbers Rebound," Dave Ress, *Richmond Times Dispatch*, May 20, 2023, 1.
8. The best way to follow these developments is to subscribe to the *Bay Journal*. See the July/August 2022 edition for excellent reporting on the crab issue by Timothy B. Wheeler and Jeremy Cox (https://www.bayjournal.com/).
9. "Menhaden Numbers Up: Coastwide Quota Increases," *Richmond Times Dispatch*, January 3, 2023.
10. A. S. Weakley, J. C. Ludwig, and J. F. Townsend, *Flora of Virginia*, ed. Bland Crowder (Richmond: Foundation of the Flora of Virginia Project, Inc., 2012).
11. See https://floraofvirginia.org/flora-app/.
12. See https://floraofvirginia.org/flora-app/.

## 21. CLIMATE CHANGE

1. Mary Rafferty, interview with author.
2. To read more about the closing of VCPC, visit https://law.wm.edu/academics/programs/jd/electives/clinics/practicum_list/vacoastal/statement/index.php.

## EPILOGUE

1. See https://fred.stlouisfed.org/series/GDP.
2. For more information about the Chesapeake Bay Funders Network, see https://www.chesbayfunders.org/.
3. See https://www.vee.org/grant-programs-application/.
4. See https://www.vee.org/40th-anniversary/.
5. See https://thejudgedocumentary.com/.
6. G. K. Chesterton, *What's Wrong with the World* (Sophia Institute Press, 2022), 39 (Kindle).
7. *A Commentary on 'Laudato Si'': Examining the Background, Contributions, Implementation, and Future of Pope Francis's Encyclical*, Kevin W. Irwin (Mahwah, NJ: Paulist Press, 2016), 95. The text of Pope Francis's encyclical, published as *Laudato si': On Care for Our Common Home* (Vatican City: Vatican Publishing House, 2015), can be read at https://www.vatican.va/content/francesco/en/encyclicals/documents/papa-francesco_20150524_enciclica-laudato-si.html.

# INDEX

*Page numbers in italics refer to illustrations.*

Trex, 49–50
2 percent solution, 149–53

United Nations Environmental Programme, 10
University of Maryland, 168
University of New Mexico, 93
University of Richmond, 75
University of Virginia, 35, 39, 54, 75, 143
Uranium Advisory Council, 129
US EPA. *See* Environmental Protection Agency (EPA; US)

Valley Conservation Council (VCC), 5–6, 120–24, 127, 140–41, 162, 187–88
Virginia Association of Environmental Educators (VAEE), 70, 160
Virginia Chesapeake Bay Plan, 92
Virginia Clean Economy Act (2020), 85
Virginia Coastal Policy Center (VCPC), 179–80
Virginia Coast Reserve, 35–37, 77
Virginia Commonwealth University, 113, 148
Virginia Conservation and Recreation Foundation, 25, 125
Virginia Conservation Network (VCN), 37, 84, 85, 131–34, 177–78, 181
Virginia Department of Environmental Quality (DEQ), 53, 68, 80, 102, 137, 160
Virginia Environmental Endowment (VEE), 181–91; board of, 14–16, 19, 29–31, 135–37, 197; case study of NPDES and, 51–53; Chesapeake Bay Program and, 88–96; defining objectives of, 34–39; Elizabeth River Project and, 103–12; environmental education funding by, 75, 154–63; on environmental laws and public interest, 79, 81, 85; establishment and early work of, 1, 13–14, 19–32, 72; funding advocacy of, 98–99; funding distributed by, 2–3, 14, 23, 31, 33; idea of, 16–18; Lynnhaven River NOW and, 112–14; perpetuity of, 32–34; poll on public

attitudes by, 5, 143–48, 191; public relations strategy of, 135–36, 138–41; *State of Our Rivers* report and, 102; VCN and, 130–33; work in Kentucky, 72–74; work in West Virginia, 72–73, 74–75
Virginia Environmental Excellence Program (VEEP), 50
Virginia Environmental Justice Collaborative, 132
*Virginia Environmental Law Journal* (formerly *Virginia Journal of Natural Resources Law*), 75
Virginia Environmental Network (VEN), 130–31, 140
Virginia Environmental Quality Act (1972), 66
Virginia Erosion and Sediment Control Act (1973), 68, 95
Virginia Farm Bureau, 116
VIRGINIAforever organization, 152, 153, 182
Virginia Foundation for the Humanities, 155
Virginia Historical Society, 185, 197
Virginia Institute of Marine Science (VIMS), 13, 103–4, 106, 107, 168–72
Virginia Junior Academy of Sciences (VJAS), 159–60
Virginia Land Conservation Fund, 6, 119, 125
Virginia Land Conservation Incentives Act, 203n3 (chap. 14)
Virginia League of Conservation Voters, 84, 150
Virginia Manufacturers Association, 49, 151
Virginia Marine Resources Commission, 166–67, 172–73
Virginia Natural Resources Leadership Institute (VNRLI), 143
Virginia Outdoors Foundation (VOF), 37, 118, 125
Virginia Pollutant Discharge Elimination System (VPDES), 53
Virginia State Chamber of Commerce, 77

Printed in the USA
CPSIA information can be obtained
at www.ICGtesting.com
CBHW032320010424
6237CB00004B/252

9 780813 951836